Heat and Moisture Transfer between Human Body and Environment

FOCUS SERIES

Series Editor Jean-Luc Achard

Heat and Moisture Transfer between Human Body and Environment

Jean-Paul Fohr

iSTE

WILEY

First published 2015 in Great Britain and the United States by ISTE Ltd and John Wiley & Sons, Inc.

ISTE Ltd
27-37 St George's Road
London SW19 4EU
UK

www.iste.co.uk

John Wiley & Sons, Inc.
111 River Street
Hoboken, NJ 07030
USA

www.wiley.com

Library of Congress Control Number: 2015951443

British Library Cataloguing-in-Publication Data
A CIP record for this book is available from the British Library
ISSN 2051-2481 (Print)
ISSN 2051-249X (Online)
ISBN 978-1-84821-893-2

Contents

Preface ix

Chapter 1. Building a Model for a Coupled Problem 1

1.1. Basic equations of the models (Appendix 1) 2
1.2. Boundary layers 3
1.2.1. Forced convection 4
1.2.2. Natural convection 6
1.3. Heat balance for a "system" and boundary conditions 8
1.4. On the problem of cooling of a cup of tea 11
1.4.1. Balance equations. 12
1.4.2. Research of transfer correlations 13
1.4.3. Surface temperature as a function of average temperature of the liquid 15
1.4.4. Liquid temperature as a function of time 16
1.5. Bather on a beach 19

Chapter 2. Approximate Determination of Transfer Coefficients 25

2.1. Natural convection around an isolated sphere. 25
2.1.1. Equations of boundary layers depending on velocity and temperature 26
2.1.2. Integration over the boundary layer thickness 28
2.1.3. Dimensionless formulation 32
2.1.4. Numerical solution 33
2.2. Coupled exchanges around the head of a baby lying down. 37
2.2.1. System of equations 38

2.2.2. Boundary layers for the horizontal disk 40
2.2.3. Boundary layers on curved surfaces 41
2.3. Forced convection around a cylinder 43
2.3.1. System of equations 44
2.3.2. Integration of the equations of the dynamic boundary layer 46
2.3.3. Dimensionless integral equation 48
2.3.4. Resolution of the upwind dynamic boundary layer 50
2.3.5. Resolution of the downwind dynamic boundary layer 55
2.3.6. Resolution of the thermal boundary layer 56

Chapter 3. Human Thermal Models 61

3.1. The Fanger model: from climatic chamber to standard 61
3.1.1. Environment and human body physical parameters 62
3.1.2. Equilibrium balance equation in the Fanger model 69
3.1.3. Examples of ambient environment qualifications 72
3.2. Gagge model 76
3.2.1. A simple, unsteady and regulated geometrical model 76
3.2.2. Response of "human system" to a sudden change in metabolism 78
3.3. Stolwijk 25 node model 80
3.4. Thermal model of a baby lying down 82
3.4.1. Geometrical division 82
3.4.2. Metabolism and respiration 83
3.4.3. Exchanges of the uncovered part of the head 84
3.4.4. Conduction between body layers 85
3.4.5. Sensible heat exchanges of the trunk 87
3.4.6. Trunk evaporation 88
3.4.7. Blood convection 89
3.4.8. System of equations 90
3.4.9. Simulation results 91

Chapter 4. Heat and Humidity Transfer in Clothing 97

4.1. From heterogeneous porous to continuous model media 98
4.2. Heat diffusion and convection 100
4.3. Vapor diffusion 101
4.4. The effect of bound water 105
4.5. Liquid water diffusion 111
4.6. Mass and energy balances 119
4.7. Limit conditions 121

4.8. Processing for a numerical resolution 123
4.9. First example: condensation in a multilayer. 124
4.10. Convection and diffusion 128
4.11. Taking account of radiation 130
4.12. Second example: firefighters' clothing 135
4.13. Traditional warm weather clothing 137

Appendices 143

Appendix 1 145

Appendix 2 151

Appendix 3 155

Bibliography 157

Index 161

Preface

The result of long experience, this work addresses audiences in the teaching and research of “fluid mechanics and heat transfer” with an introduction to physiology, as it approaches media surrounding human body. The teaching aspect involves a method of presenting concepts, models, equations and examples of application. The research aspect consists of presenting coherent physical models, such as textile layers, and revisiting old methods in fluid mechanics (boundary layer integrals) that preceded commercial computation codes, “black boxes” that were difficult to handle.

The heat and humidity exchanges between the human body and environment were approached some 50 years ago by physiologists, who were concerned about the thermal comfort and protection of the human body under difficult working conditions. Driven by entrepreneurs and manufacturers, the concept of “comfort” has in time evolved to cover all aspects of human life (home, office spaces, transport, clothing, etc.), but it took a while for more physical studies to emerge, and they are still incomplete. For example, it is very difficult to determine the radiation parameters of textile layers, which are still ignored by manufacturers.

The physical aspects of heat and humidity exchanges in this context highlight multiple facets that are often incompatible with the current narrow specialization of researchers. Let us take some examples: dynamic thermal and mass boundary layers, forced and natural convection, modeling of the “human thermal system”, heat and humidity transfers in a porous textile environment, solar and infrared radiation through porous layers, etc. To be an expert for these aspects is the result of a scientific "culture", but it requires physical good sense in order to drive a scale analysis of the equations and to interpret computing results for the selected models.

This book is composed of four chapters that reflect the learning progression of readers with a solid background in fluid mechanics and heat transfer.

Chapter 1 is a synthesis of tools for processing low-speed exchanges between a "system" and its environment, through either forced or natural convection. The examples are taken from common experience and rigorously reflect a "modeling-evaluation" approach, preliminary to a more detailed study.

Chapter 2, focuses on dynamic thermal and mass boundary layers, clarifies the common expression "layer of ignorance" taken into consideration by an exchange coefficient taken from the literature. The development of discretization methods and high-performance computation tools for a system of partial differential equations (the numerical analyst's pride) has pushed integral methods into oblivion. We can gladly revisit these when we are confident that we can afterward easily find numerical solutions to one or two integral equations. We may sometimes be want to input some experimental data (pressure curve and point of detachment) in order to deduce the exchanges around a body; the approach of such exchanges around a baby's head shows the power of the method.

Chapter 3 explores human thermal models from a historical and normative perspective. The simplest is the basis for the current standard of comfort in the habitat, while the most complex aims to understand "comfort" in the case of astronauts. The example of a baby lying down illustrates all the stages that need to be passes through in order to solve a "system's" complexity.

Chapter 4 analyses all the facets of heat exchanges (conduction, convection and radiation) and humidity exchanges (vapor, bound water and liquid water) through textile layers. It indicates how to measure key parameters of various transfer modes, whether or not in relation to the norms. A general model is then proposed as a system of two equations, mass and heat balance, a complex system which will often get simpler for applications. Highly typical cases (firefighter's garment, traditional garment of the Sahelian people, etc.) are offered as examples clarify these equations and show that physics is closely related to calculation.

J.P. FOHR
September 2015

1

Building a Model for a Coupled Problem

There are numerous and varied heat and humidity exchange coupled problems in the environment, and more specifically in man's surrounding environment (comfort, habitat, clothing, etc.) and a common methodology to approach these can be established. First of all, we need to position ourselves in relation to the digital/IT tools currently offered in "the market" and which allow for a resolution of numerous physics problems. The readers may be under the impression that the difficulty resides rather in making a choice among all these tools/software. It is common for specialized research departments to use software adapted to their fields (habitat, aviation, automobile, etc.) though a layman perceives them as some sort of magic "black box". When results come out, the reliability interval is often uncertain, as the given problem was never treated for a neighboring configuration. It should be noted that solving a mathematical model numerically with elaborated software presupposes the formulation of a number of simplifying hypotheses that may be valid for a given configuration, but risky for another. To take an extreme example, outside of our field of study, media report on the progress of IPCC works concerning climate heating predictions while they highlight the uncertainty of 20-year predictions. At planetary scale, ocean/atmosphere models are particularly complex.

Let us therefore consider a "system" whose thermal and hydric behavior in particular conditions is to be determined: an individual in a room, a manned vehicle, an incubator, a piece of sportswear, etc. It is always possible to set the proper orders of magnitude for the behavior of a system under thermal constraints by scale analysis of the equations of an adapted model and by using the theoretical and experimental data in the literature. This first model can be preliminary to the use of software that is more complex but more difficult to interpret under the relative influence of input parameters. Through several simple examples, we will examine the implementation of such models.

1.1. Basic equations of the models (Appendix 1)

A fluid medium (humid air, liquid water, etc.) put in motion by a machine (forced ventilation, pump, etc.), wind, temperature gradients (natural convection), can be described by a number of variables depending on space and time: pressure p, temperature T, velocity $\vec{V}$, density ρ, enthalpy h, etc.

The (quite) general conservation equations given here are written in condensed notation, using a pseudo vector $\vec{\nabla}$ (nabla), or gradient, which in Cartesian coordinates x, y, z can be written: $\vec{\nabla}\left(\frac{\partial}{\partial x},\frac{\partial}{\partial y},\frac{\partial}{\partial y}\right)$.

Mass conservation can be written as:

$$\frac{\partial \rho}{\partial t} + \vec{\nabla}.\rho\vec{V} = 0 \qquad [1.1a]$$

Or if we use the differential operator in the direction of movement $\frac{D}{Dt} = \frac{\partial}{\partial t} + \vec{V}.\vec{\nabla}$, we then have:

$$\frac{D\rho}{Dt} + \rho\vec{\nabla}.\vec{V} = 0 \qquad [1.1b]$$

For vapor contained in incompressible air, mass conservation is written as:

$$\frac{D\rho_v}{Dt} = \vec{\nabla}.(D_v\vec{\nabla}\rho_v) \qquad [1.2]$$

This equation is based on Fick's law of diffusion, which gives the mass diffusive flux $\vec{J_v}$ (kg/m^{2}s) of the vapor species (ρ_v) in the air (ρ): $\vec{J_v} = -D_v\vec{\nabla}\rho_v$, where D_v is the diffusion coefficient. This law is valid for humid air with $\rho_v \ll \rho$. For gas mixtures where this order of magnitude is no longer valid, a more precise law is applied.

The momentum conservation for the “volume forces” limited to gravity, for “Newtonian” fluids with constant viscosity coefficient μ is written as:

$$\rho \frac{D\vec{V}}{Dt} = -\vec{\nabla} p + \rho \vec{g} + \mu \nabla^2 \vec{V} \tag{1.3}$$

The first member is the inertia term and the second member contains pressure, gravity and viscosity terms.

The energy conservation or the first law of thermodynamics is written as:

$$\rho \frac{Dh}{Dt} = \frac{Dp}{Dt} + \vec{\nabla}.\left(\lambda \vec{\nabla} T\right) + \mu \Phi \tag{1.4}$$

The first member expresses enthalpy conservation, and the second one contains the “heat” corresponding to compression or expansion, an important quantity for certain machines, the conductive/convective transfer and the viscous dissipation. In certain cases, a radiation energy term can be added. Viscous dissipation, always positive, is expressed as a function of the derivatives of velocity components. At low velocities it is negligible. Therefore, the most common form of [1.4] applicable at low velocities is:

$$\rho c_p \left(\frac{\partial T}{\partial t} + \vec{V}.\vec{\nabla} T\right) = \vec{\nabla}.\left(\lambda \vec{\nabla} T\right) \tag{1.5}$$

We should note that equations [1.2] and [1.5] are similar in form, and this will have very practical analogical consequences.

We add to these the so-called equations of “state” of the fluid. For a perfect gas, which is the case of air at human environment temperature and atmospheric pressure, we have:

$$p = \rho r T \,;\, Dh = c_p DT \tag{1.6}$$

For a liquid at moderate pressures and temperatures we have:

$$\rho = cst \,;\; c_p = c_v = c \,;\, Dh = cDT + \frac{Dp}{\rho} \tag{1.7}$$

1.2. Boundary layers

A fluid in movement is limited by a solid wall surrounded by “boundary layers”, regions that despite their low thickness play a key role in the exchanges (of momentum, heat, etc.), and where important gradients (of velocity, temperature, etc.) develop. This concept originated in the development of aviation in the 1920s. Airplanes with airfoils that generated the best lift for a wide incidence range had to be designed.

1.2.1. *Forced convection [SCH 60]*

The typical example treated by fluid mechanics texts is that of a heated cylinder, with diameter D, placed perpendicular to a uniform fluid flow of velocity V. The development of boundary layers (in terms of velocity and temperature) starting from the stagnation point, then a detachment that generates unsteady wake behind the cylinder can be observed (Figure 1.1). Another velocity layer accompanies the wake's vortices. The thermal boundary layer, for heat conduction "across" the stream filaments, is continuous and adapts to velocity heterogeneities. The Reynolds number $Re = VD/\upsilon$, where υ is the kinematic viscosity, describes the flow regime as subcritical when the dynamic boundary layer remains laminar before detachment, or supercritical when the dynamic boundary layer transits from laminar to turbulent regime before detachment. The point of detachment moves downstream when turbulence emerges in the boundary layer. The critical Reynolds number is close to $4.5.10^5$.

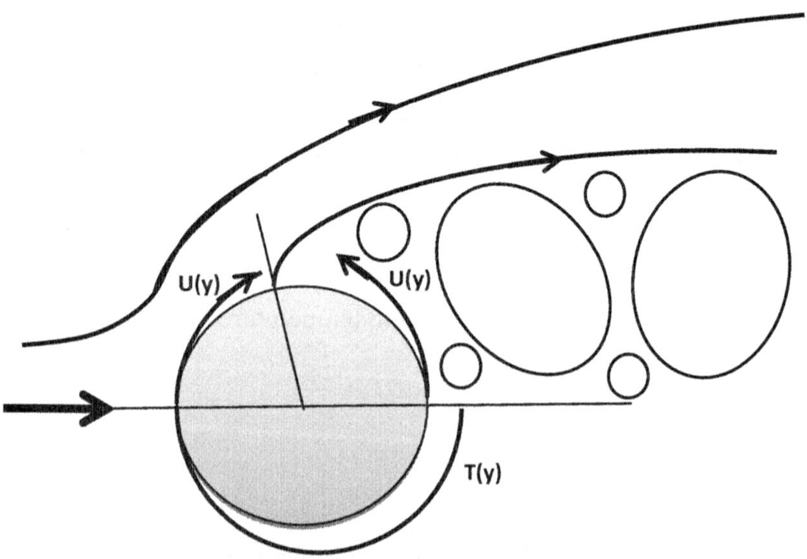

Figure 1.1. *Boundary layers around a cylinder in forced convection*

Let us suppose that this cylinder has a surface that is being heated at constant temperature and also kept humid (saturated) by an internal device. A third boundary in humidity (ρ_v) is thus established around the cylinder. The thermal boundary layers, which are key to heat exchange, are "globalized" by a heat transfer coefficient h_{cv} (not to be confused with the enthalpy h) whose definition by heat flux density is the following:

$$h_{cv}(T_p - T_o) = -\lambda \left(\frac{\partial T}{\partial n}\right)_p \quad [1.8]$$

T_p and T_o are, respectively, the temperature of the wall and of the fluid stream away from the wall, and the temperature gradient is defined at the wall along the exterior wall surface normal. For the cylinder example, h_{cv} is a function of the curvilinear abscissa. This transfer coefficient is rendered dimensionless in the form of a Nusselt number $Nu = h_{cv}L/\lambda$, where L is a length characteristic to the problem (D for a cylinder). Integrating over the surface we get an average transfer coefficient and an average Nusselt number often denoted by $\bar{h}_{cv}, \overline{Nu}$. In the case of forced convection, experimental or theoretical data take the form Nu = f (Re, Pr) which is often expressed as $Nu \approx Re^n\, Pr^p$, where Pr is Prandtl number, a characteristic of the fluid written as $Pr = \upsilon/\alpha$; $\alpha = \lambda/\rho c_p$.

Similarly, in order to characterize the "vapor" boundary layer, a mass transfer coefficient denoted by k is defined by the mass density flux:

$$k(\rho_{vp} - \rho_{vo}) = -D_v \left(\frac{\partial \rho_v}{\partial n}\right)_p \quad [1.9]$$

The dimensionless form of this coefficient is the Sherwood number $Sh = kL/D_v$, a function of Reynolds number characterizing the flow, and of Schmidt number $Sc = \upsilon/D_v$, which connects the physical properties of fluids (vapor and air in the mentioned example). In general, the data refer to surface averaged values. For a wall kept humid in such a manner that it can be considered a liquid surface, we have $\rho_{vp} = \rho_{vsat}(T_p)$, which is the condition for air to be saturated with water vapor.

In the same context of forced convection the analogy of equations [1.2], [1.5] allows us to affirm that for similar boundary conditions (constant temperature and humidity at the wall, for example) the relations Nu = f(Re, Pr) and Sh = f(Re, Sc) will be similar in form.

1.2.2. *Natural convection [BEJ 84, BEJ 93]*

Let us now consider the previous example heated cylinder in still air. Natural convection occurring around the cylinder under the influence of thermal body forces (warm air is lighter than cold air) generates a thermal boundary layer and a "plume"-like wake above it (Figure 1.2).

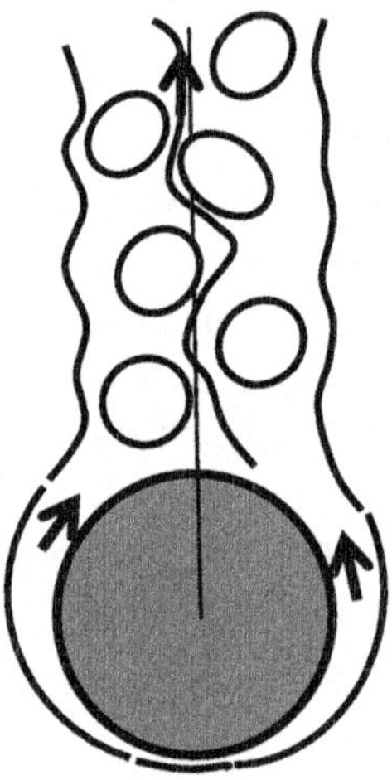

Figure 1.2. *Natural convection around a heated cylinder*

The heating intensity (wall temperature T_p) determines the nature of the flow (laminar, turbulent) and is expressed by a number, obtained by rendering the equations dimensionless, namely the Rayleigh number $Ra = \frac{g\beta(T_p - T_o)L^3}{\alpha\nu}$, where g is gravitational acceleration, $\beta = 1/T$ for a perfect gas (T Kelvin), L is a length characteristic to the problem (here, D). An alternative to Ra is the Grashof number Gr = Ra/Pr. The laminar/turbulent transition takes place around $Gr = 10^9$. The experimental correlations for various geometries of the objects on which natural convection flows occur take the form Nu = f(Ra, Pr) or Nu = f(Gr, Pr), Nusselt numbers often being averaged on the transfer surface.

Let us now complexify the previous example by supposing that the cylinder surface is not only heated, but also kept humid (a porous surface, for example). A humidity boundary layer forms (ρ_v) and the problem of vapor interfering with thermal body force emerges. The component of this force on the cylinder varies depending on the slope of the wall surface taken into consideration. To show the importance of the 2 terms of possible body force, let us consider the simple geometry of a vertical wall and write the heat transfer equations [1.3] and [1.5] projected on the 2 axes, a horizontal x axis and a vertical, ascending y axis. The respective components of velocity are u, v. The boundary layer hypotheses induce a number of simplifications u<<v, $\frac{\partial^2}{\partial x^2} \ll \frac{\partial^2}{\partial y^2}$ that lead to the system:

$$\begin{cases} -\frac{\partial p}{\partial x} = 0 \\ \rho\left(u\frac{\partial v}{\partial x} + v\frac{\partial v}{\partial y}\right) = -\rho g - \frac{\partial p}{\partial y} + \mu\left(\frac{\partial^2 v}{\partial x^2}\right) \\ u\frac{\partial T}{\partial x} + v\frac{\partial T}{\partial y} = \alpha\left(\frac{\partial^2 T}{\partial x^2}\right) \end{cases} \quad [1.10]$$

The external pressure on the boundary layer is the hydrostatic pressure p such that $p+\rho_o gy$ = cst and the system momentum conservation equation [1.10] becomes:

$$\rho\left(u\frac{\partial v}{\partial x} + v\frac{\partial v}{\partial y}\right) = -(\rho - \rho_o)g + \mu\left(\frac{\partial^2 v}{\partial x^2}\right) \quad [1.11]$$

The density of a two-component gas mixture, such as air and water vapor, is a function $\rho = \rho(T, \rho_v)$, which for small variations around a rest position T_o, ρ_o, ρ_{vo} gives:

$$\rho = \rho_o + \left(\frac{\partial \rho}{\partial T}\right)_o (T - T_o) + \left(\frac{\partial \rho}{\partial \rho_v}\right)_o (\rho_v - \rho_{vo}) \quad [1.12]$$

The derivatives of this equation are calculated starting from the humid air equations (the dry part of air will be denoted by a and the vapor part will be denoted by v):

$$\begin{cases} p = p_a + p_v \,;\, \rho = \rho_a + \rho_v \\ p_a = \rho_a \frac{R}{Mol_a} T \,;\, p_v = \rho_v \frac{R}{Mol_v} T \end{cases} \quad [1.13]$$

Mol denotes the molar mass. When $p = p_o$, equation [1.12] becomes:

$$\begin{Bmatrix} \rho = \rho_o(1 + \beta_T(T - T_o) + \beta_v(\rho_v - \rho_{vo})) \\ \beta_T = -\frac{p_o Mol_a}{R T_o^2 \rho_o} \,;\, \beta_v = \frac{1}{\rho_o}\left(1 - \frac{Mol_a}{Mol_v}\right) \end{Bmatrix} \quad [1.14]$$

The two expansive body forces terms generated by the force of gravity result from the density difference that accompanies temperature or humidity differences. Equation [1.14] shows that air density decreases when temperature increases and when humidity increases ($Mol_v/Mol_a = 0.622$). The ratio of the body forces terms indicates the relative importance of the 2 phenomena: $N = \frac{\beta_v(\rho_{vp}-\rho_{vo})}{\beta_T(T_p-T_o)}$. For humid air $N \cong 0.6\frac{\Delta\rho_v/\rho_a}{\Delta T/T_o}$ which for coupled problems results in values that are frequently $\frac{\Delta\rho_v}{\rho_a} \in (10^{-2};\ 10^{-3})$; $\frac{\Delta T}{T} \approx 10^{-1}$.

The humid body force term is usually negligible compared to the thermal one. Let us note that β_T reduces to 1/T for purely thermal natural convection.

For natural convection problems that are mainly thermal, but imply mass transfer, the similarity of equations in temperature and humidity, associated with similar boundary conditions leads to deducing the correlations Sh = f(Ra, Sc) from those of Nu = f(Ra, Pr), similarly to forced convection.

What happens to these systems of equations when, above a certain Reynolds or Rayleigh number, a part of the boundary layers, in either forced or natural convection, go through a transition to a turbulent regime? Turbulence is characterized by local velocity, pressure, temperature... fluctuations over a characteristic time that allows for a definition of average values over this period of time. The averaging procedures for the previously introduced equations lead to additional quadratic terms that we can "interpret" as friction or turbulent conduction.

A flow classification vocabulary was thus introduced: forced or natural convection, laminar or turbulent regime, to which we can add steady state (wind tunnel experiment) or unsteady state (wind on a building), internal flow (through a pipe) or external flow (around an obstacle).

1.3. Heat balance for a "system" and boundary conditions [FOH 10]

The heat balance equation for a "system" under study (a human body, a room, etc.) is the first law of thermodynamics or the energy conservation law equation, generally without the term corresponding to mechanical power exchange with a machine. For an "open" system, defined on a domain $\mathcal{D}$ that allows for incoming or outgoing mass fluxes through sections σ_i, the equation is written:

$$\begin{cases} \frac{dE}{dt} + h_i \dot{m}_i = \dot{Q} \\ E = \int_{\mathcal{D}} \rho e d\mathcal{V};\ e = u + \frac{V^2}{2} + gz \\ h = e + p/\rho \end{cases} \qquad [1.15]$$

E is the energy of the system that "receives" a heat flux $\dot{Q}$ at its boundary (transferred through radiation, convection on the wall) and eventually an internal flux of combustion heat or radiation. The notations e, u, h, ρ are relative to mass unit (or volume unit for ρ) with u as internal energy, h as enthalpy. Pressure is denoted by p, potential energy by ρgz, $\dot{m}$ is the outgoing mass flow rate from a domain, and which in forced convection is defined by $\int_{\sigma_i} \rho \vec{V}.\vec{n}_{ext} d\sigma$ and in diffusion through a boundary layer is written $\int_{\sigma_i} -D_v \frac{\partial \rho_v}{\partial n_{ext}} d\sigma$, where $\vec{n}_{ext}$ is the external normal to the boundary of the domain $\mathcal{D}$. For a perfect gas du = c_v dT, dh = c_p dT. For a liquid within a range of normal temperature and pressure $c_p = c_v$ and du = dh. Depending on the given problem, certain hypotheses will be needed: incompressibility, negligible kinetic energy, secondary gravity effect, etc.

The radiative flux at a solid surface in a gas medium, which is limited here to air, is quite complex and therefore its calculation requires more than a general approach: shape factors, surface radiative properties (emission, reflections and absorption) depending on wavelength. Within the frame of the environment described here, radiative problems are simplified as follows: possible solar radiation is calculated by several equations that take into account directional geometry and meteorological data. Long wavelength radiation (infrared) is comprehended by simplified linearization of exchanges between gray surfaces. Let $\dot{q}_{ray} d\sigma$ be the net radiative flux (emitted minus incident) from a surface dσ. Conductive heat density flux in a solid medium is generally written $\vec{q} = -\lambda \vec{\nabla} T$. Forced or natural movements driven in a fluid medium by temperature gradients generate convection boundary layers. At the solid-fluid boundary, where normal $\vec{n}$ is directed toward fluid, flux transferred on an elementary surface dσ will be expressed as $\vec{q}.\vec{n}d\sigma = -\lambda \frac{\partial T}{\partial n} d\sigma$. Given the $\dot{Q}$ definition, the following can be written:

$$\dot{Q} = -\int_{\sigma} \vec{q}.\vec{n}d\sigma - \int_{\sigma} \dot{q}_{ray}d\sigma - \int_{vol} \dot{q}_{ray}dv$$

For the calculation of radiative flux for a volume element (see Chapter 4) an adequate method, different from the one for the flux on a surface element, will be used.

The case of liquid boundaries subjected to evaporation or condensation should be carefully considered. It is a generalization of the expression written for the boundary condition on a surface. Let an elementary free surface thin layer be an evaporation ($\dot{m}_v > 0$) or condensation ($\dot{m}_v < 0$) site for a mass flux $\dot{m}_v \vec{n}_{ext} d\sigma$ undergoing phase change compensated by a transfer of thermal energy per unit of mass $\Delta h_v = h_{vap} - h_{liq}$ at a given temperature, namely that of the free surface (Figure 1.3).

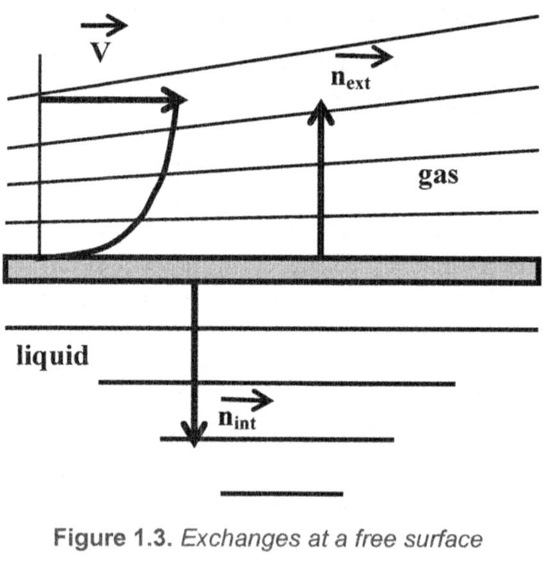

Figure 1.3. *Exchanges at a free surface*

Let $\overrightarrow{q_n}.\vec{n}d\sigma$ be the radiation heat flux along direction $\vec{n}$ from an elementary surface dσ. On each side of the surface fluids transmit energy either by conduction alone (no movement) or by convection (conduction across a boundary layer) or by radiation in a gas medium. Conduction flux density can be written as: $\dot{q_n} = -\lambda_n \frac{\partial T}{\partial n}$, where λ_n is the conductivity of medium n. The hot surface effect is denoted by $\dot{q_n} > 0$ and the cold surface effect by $\dot{q_n} < 0$.

The energy balance can thus be written (equation [1.15], E = 0):

$$\dot{m}_v \Delta h_v = -\dot{q}_{ray} - \dot{q}_{n,ext} - \dot{q}_{n,int} \tag{1.16}$$

Evaporation or condensation takes place starting from a boundary layer of vapor density ρ_v. The mass flux density is expressed as $\dot{m}_v = -D_v \frac{\partial \rho_v}{\partial n_{ext}}$ and [1.16] and can be developed as:

$$-D_v \frac{\partial \rho_v}{\partial n_{ext}} \Delta h_v = -\dot{q}_{ray} + \lambda_{ext} \frac{\partial T}{\partial n_{ext}} + \lambda_{int} \frac{\partial T}{\partial n_{int}} \tag{1.17}$$

This is a general relation applicable to both evaporation and condensation. Let us note that for each free surface problem it is possible to correctly write the boundary condition for a domain $\mathcal{D}$ precisely defined with the free surface boundary crossing either the liquid (output enthalpy $\dot{m}_v h_{liq}$, heat flux from the liquid), or the gas (output enthalpy $\dot{m}_v h_{vap}$, heat flux from the gas). Let us now apply this approach to the following example.

1.4. On the problem of cooling of a cup of tea

This example illustrates how, starting from physical reality, a model can be set up by formulating hypotheses and searching the specialist literature for transfer coefficients drawing on physical analogies.

A tea enthusiast takes the time to wait for the liquid to reach the proper temperature for tasting it. Let us consider a cup of tea filled with this drink at 90°C. We propose to predict its cooling. Let the geometry of a cylinder cup be the following: height H = 10 cm, internal diameter D = 6 cm, wall thickness ep = 5mm. The wall is made of sandstone with the following thermal characteristics: conductivity λ = 1.5 W/mK; density ρ = 2,300 kg/m^3; specific heat c_p = 710J/ kgK; emissivity ε = 0.95. The cup is set on a table in a room with a calm atmosphere, at temperature T_a = 20°C and absolute humidity W_a = 5.10^{-3}.

The inventory of thermal transfers around the cup is as follows (Figure 1.4): natural convection and radiation on the vertical side around the cup, natural convection, radiation and evaporation on the horizontal liquid surface, conduction in the cup's thickness and on the table.

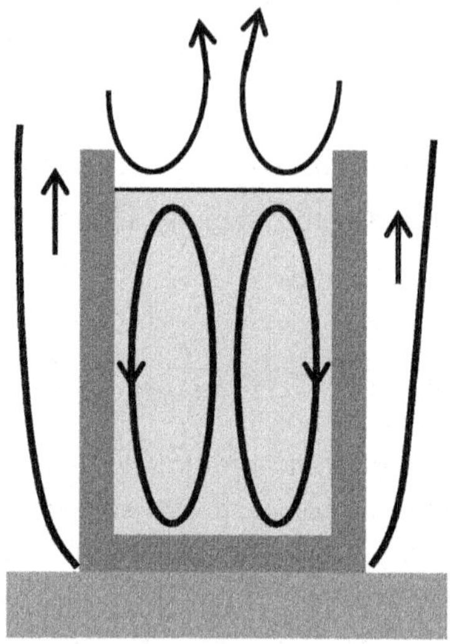

Figure 1.4. *Heat transfers around a cup of tea*

1.4.1. *Balance equations*

The heat balance equation or the first law of thermodynamics can be written on the whole for the "system" cup + liquid, the domain's boundary including the full free surface (outgoing vapor):

$$\frac{dE}{dt} + h_{vap}\dot{m}_{vap} = -\dot{Q}_{conv} - \dot{Q}_{ray} - \dot{Q}_{cond} \qquad [1.18]$$

E is the total internal energy of the filled cup. The second member is the sum of heat fluxes dissipated by convection, radiation and conduction through the table. These fluxes are algebraic quantities, but bear a negative sign because they transfer energy outside ("outgoing" fluxes). Conduction through the cup is not represented, as it is internal to the system. E decomposition can be written as:

$$E = (Mc_p\overline{T})_t + (Mc_p\overline{T})_{liq} \qquad [1.19]$$

Indices t and liq refer to the solid object cup and to the liquid. $\overline{T}$ is a spatial average temperature and M_{liq} is a function of time, as for temperatures. The derivation of E leads to:

$$(Mc_p)_t\frac{d\overline{T}_t}{dt}+(Mc_p)_{liq}\frac{d\overline{T}_{liq}}{dt}+(c_p\overline{T})_{liq}\frac{dM_{liq}}{dt}+h_{vap}\dot{m}_{evap}=$$
$$-\dot{Q}_{conv}-\dot{Q}_{ray}-\dot{Q}_{cond} \qquad [1.20]$$

Let us introduce mass conservation:

$$\frac{dM_{liq}}{dt}=-\dot{m}_{liq};\ \dot{m}_{liq}=\dot{m}_{evap} \qquad [1.21]$$

and the differential evaporation enthalpy $\Delta h_v=h_{vap}-h_{liq}$; $h_{liq}=(c_pT)_{liq}$. We obtain:

$$(Mc_p)_t\frac{d\overline{T}_t}{dt}+(Mc_p)_{liq}\frac{d\overline{T}_{liq}}{dt}=-\dot{Q}_{conv}-\dot{Q}_{ray}-\dot{Q}_{cond}-\dot{Q}_{evap} \qquad [1.22]$$

where $\dot{Q}_{evap}=\dot{m}_{evap}\Delta h_v(T_{surf})$; T_{surf} is the free surface temperature, $\dot{m}_{evap}>0$.

This equation [1.22] can be formulated directly, somewhat intuitively. We already have three temperatures $\overline{T}_t, \overline{T}_{liq}, T_{surf}$ that were introduced for one equation. It is now time to formulate the physical hypotheses and to explain the terms.

The energy balance on the free surface (equation [1.17]) gives a second equation:

$$\dot{Q}_{liq}=\dot{Q}_{conv,surf}+\dot{Q}_{ray,surf}+\dot{Q}_{evap} \qquad [1.23]$$

which expresses that the heat flux provided by the liquid mass to the surface, to be determined by internal transfers, is dissipated by evaporation, convection and radiation.

1.4.2. *Research of transfer correlations*

The temperature field is indeed dependant on the natural convection flows. Careful observation of the liquid surface leads to detecting natural convection cells that are slowly changing. These cells are related to Bénard cells, observed between two horizontal plates, the lower one being heated and the higher one cooled. We will use the data of this configuration. The vertical walls may disturb the development of convection cells, but "physical intuition" of the problem validates this choice. The Rayleigh number that determines the convective movement depends on the height H and on ΔT, the temperature difference between the plates. Literature indicates a correlation that may serve as an order of magnitude (Appendix 1):

$$\begin{cases} \overline{Nu_H} = 0.069 Ra_H^{1/3} Pr_{liq}^{0.074}\ ;\ Ra_H = \left(\frac{g\beta}{\alpha\nu}\right)_{liq} \Delta T_{liq} H^3 \\ \overline{Nu_H} = \frac{\bar{q} H}{\lambda_{liq} \Delta T_{liq}}\ ; 3.10^5 < Ra_H < 7.10^9 \end{cases} \qquad [1.24]$$

This correlation is established for a permanent exchange and we will use it to get a good order of magnitude for the slow cooling of the liquid mass. The passage from flux density on the free surface to flux on the whole surface is expressed as $\dot{Q}_{liq} = \bar{\dot{q}} \sigma_{surf}$. We set an uniform temperature gradient along the cup's vertical axis and thus an average mass temperature $\overline{T}_{liq} = T_{surf} + \frac{\Delta T_{liq}}{2}$. Let us note that [1.24] shows that the flux $\bar{\dot{q}}$ does not depend on height H.

The convective flux on the free surface can be written as:

$$\dot{Q}_{conv,surf} = h_{cv,surf}(T_{surf} - T_o)\sigma_{surf} \qquad [1.25]$$

The radiative flux on the free surface is denoted:

$$\dot{Q}_{ray,surf} = h_{ray,surf}(T_{surf} - T_o)\sigma_{surf} \qquad [1.26]$$

The evaporative flux on the free surface is expressed as:

$$\dot{Q}_{evap} = k\left(\rho_{vsat}(T_{surf}) - \rho_{vo}\right)\Delta h_v \sigma_{surf} \qquad [1.27a]$$

Or in terms of humid air data (Appendix 2), $W = \frac{\rho_v}{\rho_a}$:

$$\dot{Q}_{evap} \cong k\overline{\rho_a}\left(W_{sat}(T_{surf}) - W_o\right)\Delta h_v \sigma_{surf} \qquad [1.27b]$$

The mass transfer coefficient k “summarizes” the vapor boundary layer to the free surface, basis of the convection plume in the air. This coefficient is evaluated by analogy with heat transfer by natural convection above a hot plate. For this configuration we obtain (Appendix 1):

$$\begin{cases} \overline{Nu_L} = 0.54 Ra_L^{1/4};\ Ra_L = \left(\frac{g\beta}{\alpha\nu}\right)_{air} \Delta T_{air} L^3 \\ \overline{Nu_L} = \frac{h_{cv,surf} L}{\lambda_{air}};\ 10^4 < Ra_L < 10^7 \end{cases} \qquad [1.28]$$

The reference length L is equal to surface area divided by perimeter, which for a disk of diameter D, gives L = D/4. The temperature difference is $\Delta T_{air} = T_{surf} - T_o$. Correlation [1.28] does not depend on the Prandtl number, and it leads to:

$$\overline{Sh_L} = \overline{Nu_L} \Rightarrow \frac{\overline{k}L}{D_v} = \frac{\overline{h}_{cv,surf}}{\lambda} L \Rightarrow \overline{k} = D_v \frac{\overline{h}_{cv,surf}}{\lambda}$$

For a good order of magnitude: $D_v = 2.6.10^{-5}$ m^2/s; $\lambda = 2.6.10^{-2}$ W/mK, namely k(m/s) = $10^{-3} h_{cv,surf}$ (W/m^{2}s).

1.4.3. *Surface temperature as a function of average temperature of the liquid*

Equation [1.23] further developed becomes:

$$\begin{cases} 0.069 \lambda_{liq} \Delta T_{liq}^{4/3} \left(\frac{g\beta}{\alpha\nu}\right)_{liq}^{1/3} Pr_{liq}^{0.074} = k \overline{\rho}_a \left(W_{sat}(T_{surf}) - W_o\right) \Delta h_v + \left(h_{cv,surf} + h_{ray}\right) \Delta T_{air} \\ k = h_{cv,surf} \frac{D_v}{\lambda_{air}};\ h_{cv,surf} = 0.54 \lambda_{air} \left(\frac{g\beta}{\alpha\nu}\right)_{air}^{1/4} \left(\frac{\Delta T_{air}}{L}\right)^{1/4} \end{cases} \qquad [1.29]$$

The coefficient h_{ray} results from a linearization of a radiative flux between a “clear” surface and surrounding walls at temperature T_o:

$$\dot{q} = \epsilon\sigma\left(T_p^4 - T_o^4\right) \cong 4\epsilon\sigma\overline{T}^3 \left(T_p - T_o\right) = h_{ray}\left(T_p - T_o\right) \qquad [1.30]$$

where ε is the emissivity of the gray surface (close to unit) and σ denotes the Stefan–Boltzmann constant, with a value of $5.67.10^{-8}$ W/m^{2}K. $\overline{T}$ is an average temperature (in Kelvin) between T_p and T_o. For example, $\overline{T}$ = 293 K, ε = 0.95 give h_{ray} = 5.4 W/m^{2}K.

Equation [1.29] is in fact a function $T_{surf}(\overline{T}_{liq})$. The various physical constants introduced have to be evaluated for an average value between the 2 temperatures. Solving this equation with a little program or on a spreadsheet readily allows for physics related remarks (Figure 1.5).

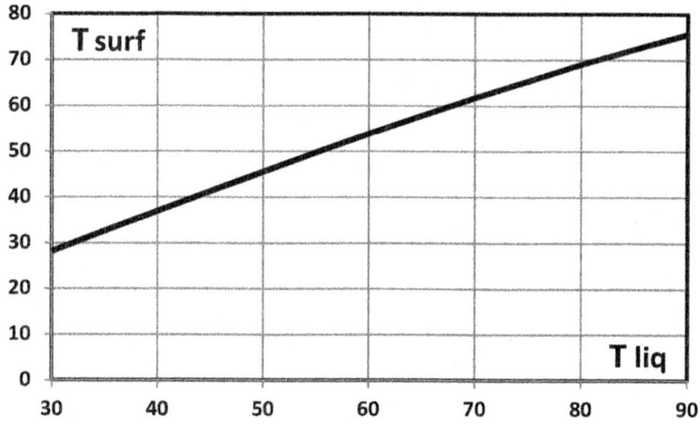

Figure 1.5. *Surface temperature as a function of the average liquid temperature*

The difference ΔT_{liq} follows a quasi-linear decrease between 30 and 90°C. The heat flux generated by the liquid mass and redistributed across the surface drastically decreases with temperature: 61.6 W when $\overline{T}_{liq}$ = 90°C and 2.4 W when $\overline{T}_{liq}$ = 30°C. The relative importance of various heat fluxes across the surface reinforces the physics intuition of phenomena.

T_{liq}	$\dot{Q}_{liq}$	$\dot{Q}_{evap}/\dot{Q}_{liq}$	$\dot{Q}_{cv,surf}/\dot{Q}_{liq}$	$\dot{Q}_{ray,surf}/\dot{Q}_{liq}$
90°C	61.1 W	0.94	0.05	0.01
30°C	2.4 W	0.84	0.11	0.05

Table 1.1. *Magnitude of heat fluxes across the liquid surface*

Surface convection and radiation come second to evaporation.

1.4.4. *Liquid temperature as a function of time*

Equation [1.22] will be solved as a differential equation in $\overline{T}_{liq}(t)$ by formulating several additional hypotheses. Conduction in the cup's thickness can be written as a one-dimensional conduction equation (thickness is small compared to diameter):

$$\rho c_p \frac{\partial T}{\partial t} = \lambda \frac{\partial^2 T}{\partial x^2} \qquad [1.31a]$$

The "time constant" deduced from orders of magnitude suggested by this equation is:

$$\tau = \frac{e_p}{\alpha} \; ; \; \alpha = \frac{\lambda}{\varrho c_p} \qquad [1.31b]$$

From the numerical data for earthenware mentioned above (λ = 1.5 W/mK; ρ = 2,300 kg/m^3, c_p = 710 J/kgK) and for a thickness e_p = 5 mm, we have: τ = 27 s. Considering that a full cup takes much longer to cool, let us formulate the hypothesis of a quasi-steady linear profile of temperature within the cup's thickness. This will allow us to write the equality of the conductive flux in the cup's thickness and the convective and radiative fluxes in the air around the cup:

$$\left(h_{cv,p} + h_{ray,p}\right)\left(T_p - T_o\right) = \frac{\lambda}{e_p}\left(\overline{T}_{liq} - T_p\right) \qquad [1.32]$$

Let us examine the literature in search of a natural correlation on a vertical wall, in the laminar case. In effect, a rapid calculation of the Rayleigh number indicates this boundary layer is laminar.

$$\overline{Nu} = 0.51 Ra^{1/4} \; ; \; \overline{Nu} = \frac{h_{cv,p} H}{\lambda_a} \; ; Ra = \left(\frac{g\beta}{\alpha\upsilon}\right)_{air} \Delta T_p H^3 \qquad [1.33]$$

where $\Delta T_p = T_p - T_o$, T_p is the temperature of the cup's external side surface. Considering [1.32] the temperature difference can be written:

$$T_p - T_o = a_p\left(\overline{T}_{liq} - T_o\right); \; a_p = \frac{\lambda_t / e_p}{h_{cv,p} + h_{ray,p} + \lambda_t / e_p} \qquad [1.34]$$

We will not take into consideration the conductive flux on the horizontal support plane, which might be notable in the case of a metal table, but negligible on a wooden plate. Equation [1.22] can then be written:

$$\left[\left(Mc_p\right)_t \frac{(1+a_p)}{2} + \left(Mc_p\right)_{liq}\right] \frac{d\overline{T}_{liq}}{dt} = -\dot{Q}_{liq} - \left(h_{cv,p} + h_{ray,p}\right)\left(T_p - T_o\right)\sigma_p \qquad [1.35]$$

To solve this last equation it is convenient to calculate numerically its second member as a function of $\overline{T}_{liq}$, and to deduce an adequate representation under the form of a second degree equation $\overline{T}_{liq} = -ax^2 + b$. In effect, coefficients $\lambda, \frac{g\beta}{\alpha\upsilon}, Pr$ are functions of temperature. M_{liq} will be considered constant in the first member, a_p practically equal to unit, $h_{ray} = 5.4$ W/m^{2}K.

The integration of $A\frac{dx}{dt} = -ax^2 + b$ gives: $t = \frac{A}{2\sqrt{ab}} Ln\left(\frac{x\sqrt{a}+\sqrt{b}}{x\sqrt{a}-\sqrt{b}}\right) + cst$ from which we deduce x(t), that is to say $\overline{T}_{liq}(t)$ (Figure 1.6).

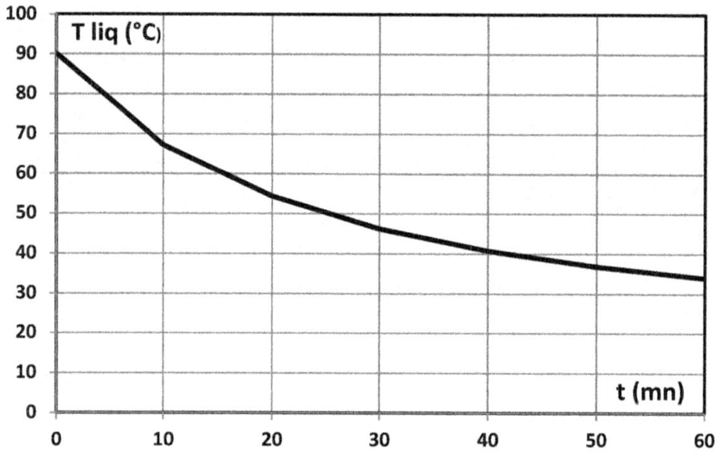

Figure 1.6. *Cooling of liquid in a cup*

We deduce from this the importance of the flux convected/radiated by the cup's vertical wall in the total flux provided by the liquid at the free surface, namely $\left(\dot{Q}_{cv,p} + \dot{Q}_{ray,p}\right)/\dot{Q}_{liq}$ which is 0.29 at 90°C and 0.88 at 30°C.

To finish the problem, let us investigate what happens when the cup reaches T_o, the external temperature. A quick examination of the subject may lead us to erroneously conclude that it reaches a state of equilibrium. Indeed, natural convection of mass takes over, as is shown by the linearization of humid air density:

$$\varrho = \varrho_o(1 + \beta_v(\varrho - \varrho_o)) \, ; \, \beta_v = \frac{1}{\varrho_o}\left(1 - \frac{Mol_a}{Mol_v}\right) \qquad [1.36]$$

The analogy between mass and heat transfer justifies the correspondences:

$$\Delta T \rightarrow \Delta\rho_v\,;\ \beta\Delta T \rightarrow \beta_v\Delta\rho_v\,;\ \alpha \rightarrow D_v\,;\ \lambda \rightarrow D_v\,; Ra \rightarrow Ra_m\,; Pr \rightarrow Sc\,; Nu \rightarrow Sh$$

We can thus transform Nu(Ra) in Sh(Ra_m) correlations. The case of same order thermal and mass body forces is more complicated and is still poorly studied. The evaporation flux requires a heat flux coming out of water or air. As soon as water temperature goes below ambient temperature by several degrees, thermal body force becomes predominant again. The N number, a ratio of the body forces terms, is equal to unit when liquid temperature is approximately 1.5°C below the ambient temperature. Equations [1.22] and [1.23] remain valid but we now have to search for correlations for the case of a free surface that is colder than the environment. Instead of [1.28] we obtain: $\overline{Nu_L} = 0.27Ra^{1/4}$ with the same definitions of Nu and Ra but $\Delta T = T_o - T_{surf}$. The same calculation method as previously, slightly simplified for "thermal equilibrium", leads to $\overline{T}_{liq} \cong 17.5°C$. In this pseudo-equilibrium case the ambient air provides a heat flux convected/radiated by the free surface and side walls. We did not take into consideration the mass reduction but it is common knowledge that a thin water layer on a plate will eventually disappear due to evaporation.

1.5. Bather on a beach

Human thermal behavior models will be examined in Chapter 3. For now, let us try to quantify the orders of magnitude of the exchanges between a naked individual on a beach and his environment, which we can compare to his own heat production (his metabolism) amounting to approximately 100 W. The bather or "beachgoer" exchanges heat through natural convection (without wind), forced convection (with sea breeze), solar radiation and infrared (long wavelength) radiation and evaporation through transpiration.

Let us evaluate convection and for this purpose let us set a simple geometrical model of a standing individual as a cylinder of height H = 1.80 m and side surface σ_p = 1.80 m², which is the skin surface area of a representative individual. We will not take into consideration in this context the exchanges through the upper base of the cylinder. The diameter of the cylinder is deduced from the above as D = 0.318 m.

Let us suppose a sea breeze with a speed V = 0.5 m/s. For the flow around the cylinder, Reynolds number Re = VD/υ is very close to 10^4 ($\upsilon = 1.6.10^{-5}$ m²/s). Among numerous correlations in the literature let us choose the following:

$$\overline{Nu} = 0.26 Re^{0.6} Pr^{0.37};\ 10^3 < Re < 2.10^5 \qquad [1.37]$$

It is an average Nusselt number integrated over the surface of the cylinder. The calculation gives (Pr = 0.72, λ = 0.026 W/mK) $\overline{Nu}$ = 57.6 and a transfer coefficient $\overline{h}_{cv} = \overline{Nu}.\frac{\lambda}{D} = 4.7\ W/m^2\ K$. The flux lost by the skin through convection can be expressed: $\dot{Q}_{cv} = \overline{h}_{cv}(T_p - T_a)\sigma_p$. Let us fix the air temperature at T_a = 25°C. The skin temperature T_p, regulated by blood flow, is in the range 31–40°C. At the latter temperature the body is near heatstroke. For T_p = 34°C we get $\dot{Q}_{cv} = 76\ W$.

Let us now suppose a calm air, with no wind, and let us consider convective exchange by natural convection. This will be determined by the Rayleigh number $Ra = \frac{g\beta}{\alpha\nu}\Delta T H^3$.

For values $\Delta T = T_p - T_a$; T_p = 34°C, T_a = 25°C, gβ/αυ = $90.7.10^6$ m³K, we have Ra = $4.7.10^9$, which would classify this flow as turbulent. Lienhard (Appendix 1) proposes the correlation:

$$\overline{Nu} = 0.52 Ra^{1/4} \qquad [1.38]$$

The calculation then gives $\overline{Nu}$ = 136; $\overline{h}_{cv}$= 6.4 W/m²K; $\dot{Q}_{cv}$= 103 W. We note that other correlations, for example that of Sparrow for a vertical cylinder with diameter equal to the height $\overline{Nu} = 0.775 Ra^{0.208}$ would give a transfer coefficient below 30% which is a common uncertainty when we manipulate literature data that were obtained in a particular context and applied to another context.

These small convection transfer calculations result in orders of magnitude that are close to those for forced and natural convection. It is a configuration of combined natural/forced convection for which few data are available.

Obviously, we could suppose that the flux exchanged will be the maximum of the two fluxes thus calculated.

The exchanges by long wavelength (infrared) radiation between skin and environment can be evaluated by [1.30] with an emissivity of skin $\varepsilon = 0.95$ or a value of the equivalent transfer coefficient $h_{ray} = 6.2$ W/m²K and an emitted flux $\dot{Q}$ of the order of 100 W.

Let us now consider the solar (short wavelengths) flux which is likely to "warm up" the skin of the "beachgoer". Everyone has experienced the extreme variability of this flux. The cylindrical geometry of the human model offers the advantage of easier calculation. Let us refer to the expression of solar radiation (Appendix 3). The question is here to evaluate a global flux received as thermal power and which depends on local meteorological data, namely global and diffuse radiation on a horizontal surface, as continuously recorded data in a weather station. Weather data are published as monthly values, but going back to the source of data allows us to get the proper orders of magnitude of clear sky radiation during summer months. Let us consider here a global solar radiation at ground level G_o between 500 and 1,000 W/m² and diffuse solar radiation D_o, which is highly dependent on the cloud cover of 30% of global radiation.

Let us suppose a surface element inclined at an angle i to the horizontal and in azimuth direction a_{surf} ($a_{surf} = 0$ to the south, positive to the west, negative to the east), the global flux received is the sum of direct and diffuse fluxes:

$$\begin{cases} G_o(i,a) = B_o R_b + D(i,a);\ B_o = G_o - D_o \\ R_b = \dfrac{cosh.sini.\cos(a_{sol} - a_{surf}) + sinh.cosi}{sinh} \\ D(i,a) = D_o\left(\dfrac{1+cosi}{2}\right) + \rho_{sol} G_o \left(\dfrac{1-cosi}{2}\right) \end{cases} \quad [1.39]$$

where h is the angular height of sun and a_{sol} its azimuth at the given moment of time, ρ_{sol} is the ground albedo, that is the reflection coefficient for solar radiation. Let us consider here $\rho_{sol} = 0.4$ which is a high value, compatible with a sand and a water surface nearby. It is important to understand that diffuse radiation comes partly from the ground and partly from the sky, solid angles being in brackets in the expression for D.

The integration over the surface of "human cylinder" is quite easy. On the whole, azimuths have no influence, and the integration gives:

$$\int_{\sigma_p} G(i,a)d\sigma = B_o \frac{2RH}{tgh} + \frac{(D_o + \rho_{sol} G_o)\sigma_p}{2}; \ \sigma_p = \pi DH \qquad [1.40]$$

Let us deduce several typical values. Let us consider the angular height of sun h = 45° in the middle of summer, during hot, rush hours on the beach. During a beautiful day values can easily reach G_o = 1,000 W/m², D_o = 300 W/m² or a flux integrated over the cylinder of 1,030 W. This is 10 times the metabolic production. Obviously, skin does not absorb all the incident solar flux, a part of it being reflected (20–30%). How does the human body react to this level of heat? Skin temperature increases and transpiration starts with a flux that, for complete skin wetness (see [1.27b]), can be expressed as:

$$\dot{Q}_{evap} = k\rho_a\left(W_{sat}(T_p) - W_a\right)\Delta h_v \sigma_p \qquad [1.41]$$

What would be the maximum value of this flux in the given context?

Let us consider the possible data: T_a = 25°C, $W_a = 7.10^{-3}$, T_p = 40°C, $W_{sat}(T_p) = 48.8.10^{-3}$, ρ_a = 1.18 kg/m³, $\Delta h_v = 2.5.10^6$ J/kgK. Because of the analogy between heat and vapor transfers in boundary layers k(m/s) = $10^{-3}h_{cv}$(W/m²s) and h_{cv} = 6W/m²K. For the "human cylinder" the calculation gives an evaporated power $\dot{Q}_{evap}$ = 1,330 W.

Heat balance is therefore often possible, but not certain, because heatstroke is also possible. On the contrary, this small calculation can explain why a bather coming out of water under a clouded sky feels cold. For example, let us consider the following fixed values: G_o = 500 W/m², D_o = 200 W/m², h = 45°, T_p = 34°C, T_a = 25°C, $W_a = 7.10^{-3}$, $W_{sat}(T_p)$ = $34.5.10^{-3}$, ρ_a = 1.18 kg/m³, k(m/s) = 10^{-3} h_{cv}(W/m²s), h_{cv} = 6 W/m²K. We get $\dot{Q}_{evap}$ = 876 W and the received solar radiation is 666 W. When evaporation, convection and long wavelength radiation are added, metabolism cannot compensate. What does the bather do? He dries in order to limit evaporation.

Going to the beach means bathing, and cooling in the water after a "grill" under the sun is a great pleasure. What is the heat loss due to sea bathing? Let us go back to the vertical motionless "cylinder/human body" model, head above water and immersed height H of 1.5 m. The order of magnitude of heat loss in water can be obtained through Lienhard correlation [1.38] provided that skin temperature is known. Immersion in water corresponds to a "thermal shock" that will make skin go from its "air" temperature (33–40°C) to a temperature close to that of water T_{eau} that we fix at 20°C. It takes several seconds for this physiological adaptation to take place, and it implies internal regulation: vasoconstriction that diverts blood from external flow (skin) to internal flow. The muscles and body core are sufficiently irrigated to distribute heat produced by metabolism, while the skin and related fat layers are "left" to their own devices, meaning that they undergo conductive heat transfers. A steady conductive regime is established in this skin/fat layer with thickness e_p within a time expressed by [1.31a], [1.31b] $\tau = e_p^2/\alpha$. Let us take the thermal data of a typical human body: λ_c = 0.5 W/mK, ρ_c = 860 kg/m^3, c_{pc} = 3,500 J/kg and thus $\alpha_c = (\lambda/\rho c_p)_c$ = 1.66.10^{-7}m²/s and if we estimate e_p between 0.5 and 1 cm, τ is approximately between 2.5 and 10 min. Beyond this period of time needed for the adaptation of the skin/fat layer we can presume a quasi-steady state of this layer and a slow cooling of the body according to the following energy equation:

$$\begin{cases} Mc_p\frac{d\overline{T}_n}{dt} = \dot{Q}_{met} - h_{cv}\left(T_p - T_{eau}\right)\sigma_p \\ h_{cv}\left(T_p - T_{eau}\right) = \frac{\lambda_c}{e_p}\left(T_n - T_p\right) \end{cases} \quad [1.42]$$

$\dot{Q}_{met}$ is the metabolic energy supplied by the body at the level of its central core supposedly at temperature (rendered uniform by blood flow) T_n. The transfer coefficient h_{cv} is obtained from [1.38]:

$$\overline{Nu} = 0.52Ra^{1/4}; \overline{Nu} = \frac{h_{cv}H}{\lambda}; Ra = \frac{g\beta}{\alpha\upsilon}\Delta TH^3; \ \Delta T = T_p - T_{eau} \quad [1.43]$$

Let us consider T_{eau} = 20°C, $T_n(t = 0)$ = 37°C, e_p = 1 cm, H = 1.5 m, D = 0.318 m, $\sigma_p = \pi DH$, M = 68.3 kg. We will not take into account the exchanges between the head above water and the air. Water physical constants are taken at an average temperature water/skin of 25°C: λ = 0.60 W/mK, $g\beta/\alpha\upsilon$ = 1.98.10^{10} m^{-3}K^{-1}.

Equations [1.42] and [1.43] are solved by iterative calculation, for example on a spreadsheet. At the beginning of cooling we get h_{cv} = 151.5 W/m²K, a convective flux lost in water $\dot{Q}_{conv}$ = 957.8 W for a motionless person metabolic flux of 100 W. A person weighing 80 kg who stands motionless for 15 min in water at 20°C would lower his body temperature by approximately 3°C.

Let us now consider the fetal position, in which losses are minimized, and assimilate it with a sphere with diameter D = 0.61 m and a volume equal to the one of the previously discussed immersed cylinder. Churchill correlation with D as characteristic length is indicated for the loss calculation made using the same method as previously:

$$\overline{Nu} = 2 + 0.589Ra^{1/4}\left[1 + (\frac{0.469}{Pr})^{9/16}\right]^{-4/9}; 1 < Ra < 10^{11} \qquad [1.44]$$

For water at this temperature scale, Pr = 6.21. We get h_{cv} = 189.5 W/m²K and a convective flux lost in water $\dot{Q}_{conv}$ = 787 W.

Let us now consider the swimmer is moving at a speed V = 1 m/s. Careful observation of the flow around the swimmer shows that stream filaments around the body cover a distance L of approximately 50 cm before being ejected sideways. It seems possible to use for an order of magnitude a “flat plate” of laminar or turbulent flow correlation, and the flow characteristic Reynolds number is here Re = VL/υ which is $5.6.10^5$, which indicates the beginning of a laminar/turbulent transition and Nusselt number is then a solution of:

$$\overline{Nu} = \frac{h_{cv}L}{\upsilon} = 0.037\left(Re^{\frac{4}{5}} - 23{,}550\right)Pr^{\frac{1}{3}}; Re > 5.10^5 \qquad [1.45]$$

We then get: h_{cv} = 1,315 W/m²K and $\dot{Q}_{conv}$ = 1,227 W. The metabolic flux that results from physical effort is here far higher than 100 W. A possible flux for an energetic swim at 600 W leads to a cooling of the said swimmer body by 2° within 15 min.

This highly simplified human body behavior model gives orders of magnitude which are quite satisfactory. In Chapter 3, we will take a look at more elaborate models.

2

Approximate Determination of Transfer Coefficients

Satisfactory expressions for heat and mass transfer coefficients for each specific configuration are not always available in the scientific literature. Resorting to heat transfer specialized software may certainly offer great opportunities, but requires significant expertise in mastering these codes and the proper evaluation of the physical choices introduced. The methods for approximate calculation of the boundary layers are still relatively easy to resort to, provided that the "philosophy" is well understood and there is a physical grasp of the phenomenon. The problem/example presented further on consists of the determination of transfer coefficients around the head of a baby who is lying down, knowing that, as we will see in the next chapter, in "overheating" situations, it is a primary factor for heat balance evaluation.

2.1. Natural convection around an isolated sphere

Let us consider in a first stage a free of contact sphere with diameter D = 2R, heated at uniform wall temperature T_p in the air at temperature T_o ($\Delta T = T_p - T_o$). Heat engineering literature presents various correlations for this case, and among them the Churchill correlation (Appendix 1) is often proposed:

$$\overline{Nu} = 2 + 0.589 Ra^{\frac{1}{4}} \left[1 + (0.469/Pr)^{\frac{9}{16}}\right]^{-4/9} ; Ra = \frac{g\beta\Delta T D^3}{\alpha \upsilon}; Pr = \frac{\upsilon}{\alpha} \quad [2.1]$$

For $Ra < 10^{11}$; $Pr \geq 0.7$.

Let us note that laminar/turbulent transition (10^8-10^9) which should take place within the validity range of Rayleigh number does not seem to disturb the correlation.

We will try to find a value equivalent to this correlation by an approximate method, named the integral method, for the calculation of laminar boundary layers. The natural convection flow around a heated sphere does not *a priori* present detachment of the boundary layer, at least not at the considered Rayleigh number of approximately 5.10^6 (D = 13 cm, $\Delta T = 20$ K, Pr = 0.71) which corresponds to laminar flow.

2.1.1. *Equations of boundary layers depending on velocity and temperature*

Integral approximate methods are based on the principle of solving the integral over the boundary layer of the mass, momentum and energy conservation equations, starting from an *a priori* set velocity and temperature profile. These methods lead to satisfactory results if they are supported by experimental data and if profiles are representative. Nevertheless, it has to be stressed that boundary layer equations are derived starting from complete equations simplifications whose validity can be challenged in specific regions: boundary layer initiation point, detachment, wake... For the subject at hand, a sphere under natural convection, the boundary layers feed a hot plume that rises above, and the validity of the calculation at the top of the sphere can be called into question.

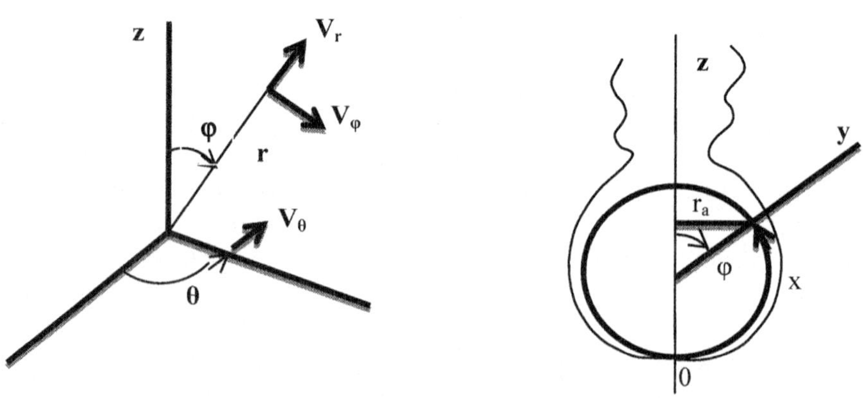

Figure 2.1. *Spherical coordinates*

The boundary layer equations in spherical coordinates provided by heat transfer books [BEJ 93] are voluminous, so we are going to use them in an already simplified form. Rotational symmetry implies: $\frac{\partial}{\partial\theta} = 0; v_\theta = 0$ for a velocity vector $\vec{V}(v_\theta, v_r, v_\varphi)$ (Figure 2.1). Unsteady terms are not taken into account.

The second transformation of the system of equations relates to the introduction of "boundary layer" variables that will lend themselves to approximations. The r, φ coordinate system is replaced by a curvilinear abscissa x on the sphere of radius R, and its origin is the lower point of the sphere, so that x = R(π-φ) and y is the distance to the sphere on the radius r = R+y. The components of velocity vector are noted u = $-v_\varphi$, v = v_r. Finally, for convenience reasons we will consider Rsinφ(x) = $r_a(x)$. Once these changes are implemented, the system of equations is as follows.

Mass conservation:

$$\frac{\partial u}{\partial x} + \left(1 + \frac{y}{R}\right)\frac{\partial v}{\partial y} + 2\frac{v}{R} + \frac{u}{r_a}\frac{dr_a}{dx} = 0 \qquad [2.2]$$

Conservation of the momentum projection on the tangent vector $\vec{x}$ (μ is constant):

$$\begin{cases} \rho\left(\frac{1}{1+\frac{y}{R}}u\frac{\partial u}{\partial x} + v\frac{\partial u}{\partial y} + \frac{uv}{R+y}\right) = -\rho g sin\varphi - \frac{1}{1+\frac{y}{R}}\frac{\partial p}{\partial x} + \\ \mu\left[\frac{\partial^2 u}{\partial y^2} + \frac{2}{R+y}\frac{\partial u}{\partial y} + \frac{1}{(1+\frac{y}{R})^2}\frac{1}{r_a}\frac{\partial}{\partial x}\left(r_a\frac{\partial u}{\partial y}\right) + \frac{2R}{(R+y)^2}\frac{\partial v}{\partial x} - \frac{1}{(1+\frac{y}{R})^2}\frac{u}{r_a^2}\right] \end{cases} \qquad [2.3]$$

The equation of conservation of momentum projection on the $\vec{y}$ axis will be discussed further on.

Viscous dissipation at low velocity is neglected and λ, c_p coefficients are constant so the energy conservation is:

$$\rho c_p\left(\frac{u}{1+y/R}\frac{\partial T}{\partial x} + v\frac{\partial T}{\partial y}\right) = \lambda\left[\frac{\partial^2 T}{\partial y^2} + \frac{2}{R+y}\frac{\partial T}{\partial y} + \frac{1}{\left(1+\frac{y}{R}\right)^2}\frac{1}{r_a}\frac{\partial}{\partial x}\left(r_a\frac{\partial T}{\partial x}\right)\right] \qquad [2.4]$$

The boundary layer equations can be simplified based on the evaluation of the order of magnitude of the terms, taking into consideration that:

$$y \ll R; v \ll u; \frac{\partial u}{\partial x} \approx \frac{u}{R}; \frac{\partial u}{\partial y} \approx \frac{u}{\delta}; \frac{\partial v}{\partial x} \approx \frac{v}{R}; \frac{\partial v}{\partial y} \approx \frac{v}{\delta}; \frac{\partial T}{\partial x} \approx \frac{\Delta T}{R}; \frac{\partial T}{\partial y} \approx \frac{\Delta T}{\delta}$$

δ is the thickness of the (dynamic or thermal) boundary layer so that $\delta << R$.

After neglecting secondary terms, the system becomes:

$$\begin{cases} \frac{\partial u}{\partial x} + \frac{\partial v}{\partial y} + \frac{u}{r_a}\frac{dr_a}{dx} = 0 \\ \rho\left(u\frac{\partial u}{\partial x} + v\frac{\partial u}{\partial y}\right) = -\rho g sin\varphi - \frac{\partial p}{\partial x} + \mu\left(\frac{\partial^2 u}{\partial y^2}\right) \\ \rho c_p\left(u\frac{\partial T}{\partial x} + v\frac{\partial T}{\partial y}\right) = \lambda\left(\frac{\partial^2 T}{\partial y^2}\right) \end{cases} \quad [2.5]$$

To the above equations we should add the equation of momentum projected on the $\vec{y}$ axis, which, considering the orders of magnitude, will retain only the pressure and gravity terms:

$$-\rho g cos\varphi - \frac{\partial p}{\partial y} = 0 \quad [2.6]$$

2.1.2. *Integration over the boundary layer thickness*

The next stage is to integrate this system of equations along y, that is to say along the thickness of the boundary layer. For rigor's sake we have to consider dynamic and thermal boundary layers whose relative importance depends on physical characteristics of the fluid *via* Prandtl number $Pr = \frac{\mu c_p}{\lambda} = \frac{v}{\alpha}$. When Pr>>1 the viscous effect predominates over the thermal conduction effect and conversely, when Pr<<1. For air (Pr = 0.7) thicknesses are close enough to be considered equal. Experimental data show profiles that tend asymptotically to zero at the border of the boundary layers, so whose thicknesses δ are not highly accurate, fact which supports this hypothesis. We will thus consider profiles that are close to experience under polynomial form (Figure 2.2).

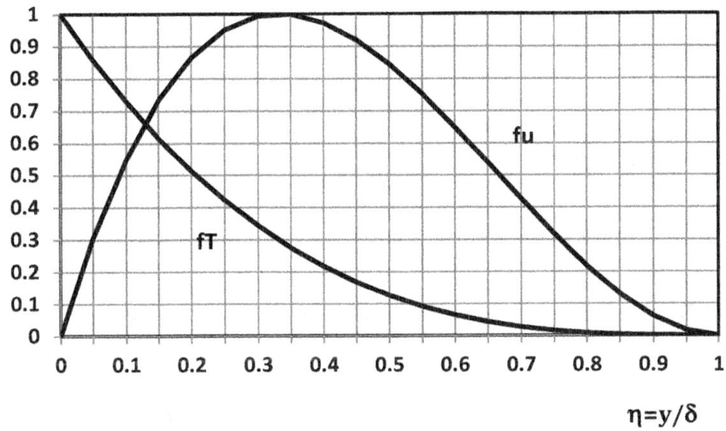

Figure 2.2. *Velocity and temperature profiles in the boundary layer*

$$\begin{cases} \frac{u}{U} = f_u(\eta); \frac{T-T_o}{\Delta T} = f_T(\eta); \eta = \frac{y}{\delta} \\ f_u(\eta) = \frac{27}{4}\eta(1-\eta)^2; f_T(\eta) = (1-\eta)^3 \end{cases} \qquad [2.7]$$

These profiles verify the following conditions at the wall and at the border of the boundary layers:

$$\begin{cases} y = 0 \Rightarrow u = 0; T = T_p \\ y = \delta \Rightarrow u = 0; T = T_o; \frac{\partial u}{\partial y} = 0; \frac{\partial T}{\partial y} = 0 \\ u_{max} = U \Rightarrow \frac{\partial u}{\partial y} = 0; \eta = 1/3 \end{cases} \qquad [2.8]$$

Prior to integration a reduction of number of variables (u, v, T, p, ρ) and equations of the system takes place. Pressure p(x, y) is derived from the equation of momentum projection along the y axis [2.6] and introduced in the equation along the x axis. We make a change of variable for the pressure in order to eliminate the static part $p = p_o + \rho_o gz$ which reigns outside the boundary layers:

$$p_d(x,y) = p + \rho_o gz - p_o \, ; \; z = R\left(1 + cos\varphi\left(1 + \frac{y}{R}\right)\right) \qquad [2.9]$$

The origin of the vertical axis is the lower point of the sphere. Considering that y/R<<1 the momentum equation along y (equation [2.6]) can be written:

$$(\rho_o - \rho)gcos\varphi - \frac{\partial p_d}{\partial y} = 0 \quad [2.10]$$

In natural convection we consider a variable density ρ only in the driving term (buoyancy) resulting from gravity terms. The linearization of the equation of state for air, considered a perfect gas, leads to:

$$\rho_o - \rho = \rho_o\beta(T - T_o);\ \beta = 1/T \quad [2.11]$$

Equation [2.10] can thus be written:

$$\frac{\partial p_d}{\partial y} = \rho_o g\beta(T - T_o)cos\varphi \quad [2.12]$$

Similarly, the 2 terms of pressure/gravity in the equation of momentum along x become:

$$\rho_o g\beta(T - T_o)sin\varphi - \frac{\partial p_d}{\partial x} \quad [2.13]$$

The elimination of variable v in [2.5] results from a combination of momentum and energy equations with the equation for mass, as follows:

$$v\frac{\partial u}{\partial y} = \frac{\partial uv}{\partial y} - u\frac{\partial v}{\partial y} = \frac{\partial uv}{\partial y} + u\left(\frac{\partial u}{\partial x} + \frac{u}{r_a}\frac{dr_a}{dx}\right) \quad [2.14]$$

The integration along the boundary layer thickness is now possible if we insert the above mentioned profiles. For the sake of reasoning rigor we should note that we are integrating between to two limits that do not depend on x, 0 and Y>δ so that we can invert the integral along y and the derivative along x. The passing to the limit Y→δ is a natural consequence of the fact that integrable functions are zero outside the boundary layer.

The first stage is the integration of [2.12] taking into account $p_d(\delta) = 0$:

$$p_d = -\rho_o g\Delta T\delta cos\varphi F(\eta);\ F(\eta) = \int_\eta^1 f_T(\eta)d\eta \quad [2.15]$$

The integration of the second equation of [2.5] considering [2.13] gives:

$$\rho_o\left[a_1\left(\frac{d}{dx}U^2\delta+\frac{U^2\delta}{r_a}\frac{dr_a}{dx}\right)\right]=\rho_o g\beta\Delta T\delta$$
$$\left[a_2 sin\varphi+a_3\frac{d}{dx}(\delta cos\varphi)+a_4\frac{d\delta}{dx}cos\varphi\right]-\mu\frac{U}{\delta}f'_u(0) \qquad [2.16]$$

where:

$$a_1=\int_0^1 f_u^2(\eta)d\eta\ ;\ a_2=\int_0^1 f_T(\eta)d\eta\ ;\ a_3=\int_0^1 F(\eta)d\eta\ ;\ a_4=\int_0^1 \eta f_T(\eta)d\eta \qquad [2.17]$$

Before the integration along y, dynamic pressure p_d is to be derived with respect to x through δ(x), φ(x), $\eta = y/\delta(x)$.

The integration of the energy equation transformed through the change of variable $T_+ = (T-T_o)/\Delta T$ taking into account the mass conservation becomes:

$$\frac{\partial uT_+}{\partial x}+\frac{\partial vT_+}{\partial y}+T_+\frac{u}{r_a}\frac{dr_a}{dx}=\alpha\frac{\partial^2 T_+}{\partial y^2} \qquad [2.18]$$

and is integrated in:

$$a_5\left(\frac{d}{dx}U\delta+\frac{U\delta}{r_a}\frac{dr_a}{dx}\right)=-\alpha\frac{f'_T(0)}{\delta}\ ;\ a_5=\int_0^1 f_u(\eta)f_T(\eta)d\eta \qquad [2.19]$$

We now have a system of two equations [2.16] and [2.19] with two unknown functions U(x) and δ(x). With the boundary layer profiles [2.7], it gives:

$a_1 = (27/4)^2/105$; $a_2 = 1/4$; $a_3 = 1/20$; $a_4 = 1/20$; $a_5 = 9/56$; $f'_u(0) = 27/4$; $f'_T(0) = -3$.

The previous system [2.16], [2.19] can be subjected to a more compact writing if the first members are transformed:

$$\begin{cases} r_a\frac{d}{dx}U^2\delta+U^2\delta\frac{dr_a}{dx}=\frac{d}{dx}(r_aU^2\delta) \\ r_a\frac{d}{dx}U\delta+U\delta\frac{dr_a}{dx}=\frac{d}{dx}(r_aU\delta) \end{cases}$$

On the other hand, the buoyancy term in the right member of [2.16] can be decomposed as follows:

$$\rho_o g\beta\Delta T\delta\left[\left(a_2 + a_3\frac{\delta}{R}\right)sin\varphi + (a_3 + a_4)cos\varphi\frac{d\delta}{dx}\right]$$

Taking into account that δ<<R, the term $a_3\delta/R$ is negligible compared to a_2. The system [2.16], [2.19] becomes:

$$\rho_o a_1\frac{d}{dx}(r_a U^2\delta) = \rho_o g\beta\Delta T r_a\delta\left[a_2 sin\varphi + (a_3 + a_4)cos\varphi\frac{d\delta}{dx}\right] - \mu r_a\frac{U}{\delta}f'_u(0) \qquad [2.20]$$

$$a_5\frac{d}{dx}(r_a U\delta) = -\alpha r_a\frac{f'_T(0)}{\delta} \qquad [2.21]$$

2.1.3. *Dimensionless formulation*

For good readability and in order to facilitate the numerical resolution of the system (terms approximately equal to unit), it is advisable to render the equations dimensionless starting from reference quantities that are characteristic to the problem. If the reference length for x can be the sphere diameter D, the boundary layer thickness has to be related to another length derived after examining the terms of the system. Let us use the classical methodology for evaluating the terms of equations [BEJ 84].

For a Prandtl number of approximately one unit the dynamic and thermal boundary layers have the same thickness δ. The mass conservation equation gives: $\frac{u}{D} \approx \frac{v}{\delta}$. The equation of motion consists of three terms of the same order: inertia in the first member and buoyancy, viscosity in the second member. The relation buoyancy $\approx$ inertia gives $\rho_o\frac{u^2}{D} \approx \rho_o g\beta\Delta T$ or a reference velocity:

$$u = \frac{\alpha}{D}(PrRa)^{1/2}; Ra = \frac{g\beta\Delta T D^3}{\alpha v}; Pr = \frac{v}{\alpha}$$

The relation inertia $\approx$ viscosity gives: $\frac{u}{D} \approx \frac{\alpha}{\delta^2}$ or in combination with the previous relation:

$$\delta \approx D(PrRa)^{-1/4}$$

We will therefore set a reference velocity $V_{ref} = \frac{\alpha}{D} Ra^{1/2}$ and a reference length for the boundary layer thickness $e_{ref} = DRa^{-1/4}$. The dimensionless variables are then:

$$U_+ = \frac{U}{V_{ref}} \; ; \; x_+ = \frac{x}{D} \; ; \; \delta_+ = \frac{\delta}{e_{ref}}$$

Equations [2.20] and [2.21] become:

$$a_1 \frac{\delta_+}{r_{a+} U_+} \frac{d}{dx_+} (r_{a+} U_+^2 \delta_+) = Pr \frac{\delta_+^2}{U_+} \left[a_2 sin\varphi + (a_3 + a_4) Ra^{-\frac{1}{4}} cos\varphi \frac{d\delta_+}{dx_+} \right]$$
$$-Prf'_u(0) \qquad [2.22]$$

$$a_5 \frac{d}{dx_+} (r_{a+} U_+ \delta_+) = \frac{r_{a+}}{\delta_+} (-f'_T(0)) \qquad [2.23]$$

From the definition of r_a we deduce:

$$r_{a+} = \frac{1}{2} sin\varphi; \; \varphi = \pi - 2x_+ \qquad [2.24]$$

New functions seem to become prominent:

$$X(x_+) = r_{a+} U_+^2 \delta_+ \; ; Y(x_+) = r_{a+} U_+ \delta_+ \Rightarrow U_+ = \frac{X}{Y} \; ; \; \delta_+ = \frac{1}{r_{a+}} \frac{Y^2}{X} \qquad [2.25]$$

However, in the buoyancy term (square brackets in the second member of [2.22]) the derivative of δ_+ seems little compatible with the new variables. Let us note that this derivative concerns a complementary term of buoyancy. In effect, it disappears for a vertical wall ($\varphi = \pi/2$), reinforces the buoyancy for the upper half of the sphere ($\varphi<\pi/2$), diminishes buoyancy for the lower half of the sphere ($\varphi>\pi/2$). It can be considered as a coefficient to be adjusted at each step of the calculation, and this saves the effort of developing the first members of [2.22] and [2.23]. We will see further on the alternative to development.

2.1.4. *Numerical solution*

In order to numerically solve the equations we have to get out of the initial singularity: $\varphi = \pi$, $x = 0$, $r_a = 0$, $U = 0$, $X = 0$, $Y = 0$. Let us examine the orders of magnitude of the terms of equations [2.22], [2.23] in the vicinity of $x = 0$, the viscosity term being of the order of one.

$$inertia \approx viscosity \Rightarrow \frac{\delta_+}{r_{a+}U_+}\frac{d}{dx_+}(r_{a+}U_+^2\delta_+) \approx \frac{\delta_+^2 U_+}{x_+} \approx 1$$

The buoyancy term consists of two parts, a driving part in sinφ (zero at the point of departure) and a "brake" part in cosφ. The order of magnitude which retains only the driving term is thus:

$$buoyancy \approx viscosity \Rightarrow \frac{\delta_+^2}{U_+}sin\varphi \approx \frac{\delta_+^2}{U_+}x_+ \approx 1$$

Furthermore, the energy equation gives: $\frac{\delta_+^2 U_+}{x_+} \approx 1$, which is similar to the information in the equation of movement. The combination of these 2 order of magnitude results leads to $\delta_+ \approx 1$; $U_+ \approx x_+$. We will thus consider in the vicinity of x = 0: $\delta_+ = \delta_o$; $U_+ = cx_+$. Let us note that the "brake" term of the buoyancy is also cancelled out. δ_o and c are deduced there from. Equation of movement [2.22] gives:

$$\left(2Pr\frac{a_2}{a_1}\frac{1}{c} - 3c\right)\delta_o^2 = Pr\frac{f'_u(0)}{a_1}$$

The energy equation gives:

$$c\delta_o^2 = \frac{-f'_T(0)}{2a_5}$$

The combination of the 2 previous relations gives:

$$c^2\left(3 + Pr\frac{2a_5}{a_1}\frac{f'_u(0)}{f'_T(0)}\right) = 2Pr\frac{a_2}{a_1} \qquad [2.26]$$

The calculation of c and δ_o then gives c = 0.444; δ_o = 4.582.

Setting a non-zero value of the boundary layer in the vicinity of a lower point of initiation of the boundary layer is quite natural, provided that thermal conduction takes place at zero velocity. Let us note that the integral equations for boundary layers are not adapted to the vicinity of the point of departure (v<<u is not respected). Having a good order of magnitude in order to start the numerical calculation is a facility that does not spare us from exploring the vicinity. In effect, the calculation convergence on the physically coherent solution may be highly sensitive to the point of departure. We will admit the solution uniqueness.

It is therefore possible to solve the system [2.22], [2.23] by correcting at each time the derivative of the boundary layer thickness or to develop this system in U_+ and δ_+ as follows:

$$\left(U_+ - a_6\frac{\delta_+}{U_+}cos\varphi\right)\frac{d\delta_+}{dx_+} + 2\delta_+\frac{dU_+}{dx_+} = -U_+\delta_+\frac{1}{r_{a+}}\frac{dr_{a+}}{dx_+} + a_7\frac{\delta_+}{U_+}sin\varphi - \frac{a_8}{\delta_+} \quad [2.27]$$

$$U_+\frac{d\delta_+}{dx_+} + \delta_+\frac{dU_+}{dx_+} = \frac{a_9}{\delta_+} - U_+\delta_+\frac{1}{r_{a+}}\frac{dr_{a+}}{dx_+} \quad [2.28]$$

where:

$$a_6^{\cdot} = \frac{(a_3+a_4)}{a_1}PrRa^{-1/4}\,;\; a_7 = \frac{a_2}{a_1}\mathrm{Pr};\, a_8 = \frac{f'_u(0)}{a_1}Pr;\; a_9 = \frac{-f'_T(0)}{a_5} \quad [2.29]$$

It is possible to get a slightly simpler system, consisting of a first equation [2.30] derived by subtracting [2.28] from [2.27], associated with [2.28]:

$$-a_6\frac{\delta_+}{U_+}cos\varphi\frac{d\delta_+}{dx_+} + \delta_+\frac{dU_+}{dx_+} = a_7\frac{\delta_+}{U_+}sin\varphi - \frac{(a_8+a_9)}{\delta_+} \quad [2.30]$$

This system of two equations with two unknowns $(\frac{d\delta_+}{dx_+}, \frac{dU_+}{dx_+})$ allows for all the numerical solutions.

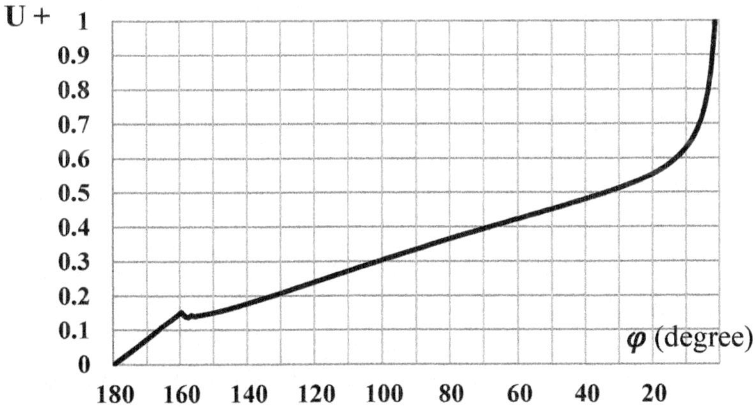

Figure 2.3. *Maximum normed velocity of the natural convection boundary layer*

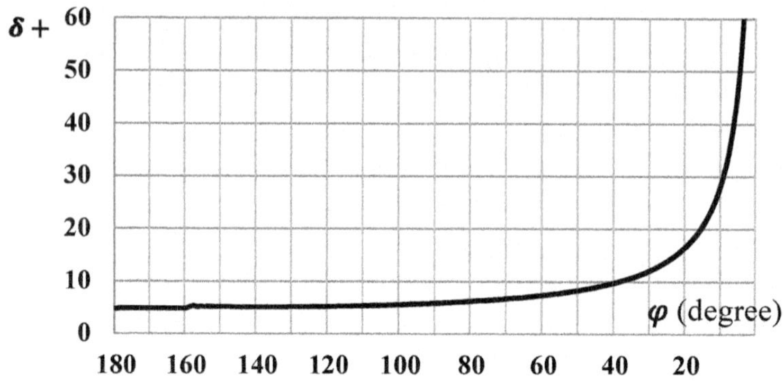

Figure 2.4. *Normed thickness of dynamic and thermal boundary layer*

Once the system is chosen, various methods are possible for its numerical resolution: homemade programming with a common language, a "turnkey" software with preset programming, a calculation set on a spreadsheet, as we have done here. The results have to be in the range with a precision of about 1 to 2%. The evolution of velocity and thickness of the boundary layer is given by Figures 2.3 and 2.4.

We note the connection of the approximate solution in the vicinity of the origin with the general solution ($\varphi = 160°$). The average Nusselt number which will allow evaluating the relevance of the calculation can be derived as follows:

$$\begin{cases} \overline{Nu} = \frac{\bar{h}D}{\lambda} \ ; \ \bar{h}\sigma_{sph} = \int_{sphère} h(x)d\sigma \\ h(T_p - T_o) = -\lambda\left(\frac{\partial T}{\partial y}\right)_p = -\lambda\frac{\Delta T}{\delta}\left(\frac{\partial T_+}{\partial \eta}\right)_{\eta=0} = 3\lambda\frac{\Delta T}{\delta} \\ \sigma_{sph} = 4\pi R^2 \ ; d\sigma = 2\pi r_a dx \ ; \ r_a = r_{a+}D \ ; x = x_+ D \ ; \ \delta = \delta_+ D Ra^{-1/4} \end{cases} \quad [2.31]$$

or:

$$\overline{Nu} = 6Ra^{1/4}\int_{sphère}\frac{r_{a+}}{\delta_+}dx_+ \quad [2.32]$$

The numerical data of the integration lead to $\overline{Nu} = 0.48Ra^{1/4}$, to be compared with the Churchill correlation (annex 1) for Pr = 0.72 or $\overline{Nu} = 0.455Ra^{1/4}$ which is a good validation of the integral method. We should note that the coefficient with a factor of $Ra^{1/4}$ depends here to a little extent on Ra by the intermediary of the "buoyancy" term in equation [2.22] and it was derived for $Ra = 5.10^6$.

Before approaching a more complex problem, let us consider an isolated sphere, similarly heated at a uniform surface temperature T_p but moreover maintained humid (with water or another liquid that evaporates easily in the air). A third boundary layer in humidity would be established, and it would be measured by the vapor density ρ_v for which the sphere surface condition would be: $\rho_v(\text{wall}) = \rho_{vsat}(T_p)$. The equations for laminar boundary layers in temperature and humidity are similar and the boundary conditions are of the same type with buoyancy exclusively thermal. A correspondence of solutions is established:

$$Nu = f(Ra, Pr) \Rightarrow Sh = f(Ra, Sc)$$

2.2. Coupled exchanges around the head of a baby lying down

The problem, which will be discussed in the next chapter, is to characterize the heat and mass (evaporation) transfers around the head of a lying down baby. In effect, excessive insulation of a baby seems to be the most important cause of sudden death, the head being the main source of heat dissipation. Let us therefore consider the following geometry of the model adopted for the representation of a head plunged into a pillow or mattress (Figure 2.5).

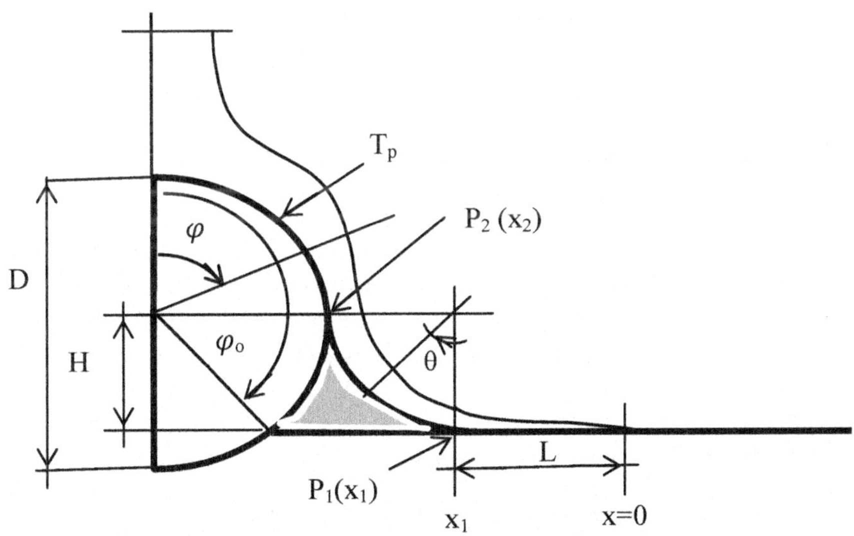

Figure 2.5. *Geometry of the baby head model – a cross-section*

The head/sphere is immersed in the support plane of a surface defined by the angle $\varphi_o = 110°$, which corresponds to a surface "in the air" of 2/3 of the sphere surface. The natural convection boundary layer that sets around the free part of the sphere is initiated starting from a planar disk on the support heated by conduction. Laboratory measurements on a physical maquette resulted in setting the dimension of this horizontal "hot spot" as the distance L defined starting from a "coupling profile" point P_1. In effect, the method of integral solutions is incompatible with a "corner" space at the base of the free part of the sphere. Based on physical considerations and observations on the maquette, we can say that the natural convection plume draws up the horizontal boundary layer on the support and induces an annular "vortex" regularly ejected in the boundary layer. The plausible model is that of a profile coupled to the base of the sphere by an arc of circle of radius H tangent to the sphere (at point P_2) and to the plane (at point P_1).

Let us notate $H = R\cos(\pi-\varphi_o)$. On this toroidal coupling profile the radius vector previously defined will be $r_a = R+H(1-\sin\theta)$. For this coupling and for the horizontal ring we will impose the same boundary condition in velocity as for the sphere: u = 0, v = 0, but with a gradual rise in temperature starting from x = 0, and equally in water vapor starting from $x = x_1$ according to the expression:

$$\begin{cases} f_{Tp}(x) = \frac{T_p(x)-T_o}{\Delta T} = \left(\frac{x}{x_2}\right)^2 \; ; 0 < x < x_2 \; ; \; \Delta T = T_{po} - T_o \\ f_{cp}(x) = \frac{\rho_{vp}(x)-\rho_{vo}}{\Delta\rho_v} = \left(\frac{x-x_1}{x_2-x_1}\right)^2 \; ; \; x_1 < x < x_2 \; ; \; \Delta\rho_v = \rho_{vsat}(T_{po}) - \rho_{vo} \end{cases} \quad [2.33]$$

T_{po} is the temperature of the upper half of the sphere surface.

This form results from the examination of the experimental data obtained on the physical maquette. Let us note that the "model" does not transpire on the coupling or on the planar disk, but is recovering water vapor from the open corner cavity. A model should be as close as possible to the physical conditions, while introducing simplifications to be discussed.

2.2.1. *System of equations*

The problem under consideration is now that of coupled heat and humidity transfers. In effect, the overheated baby transpires abundantly and we will therefore place an additional boundary condition on the free sphere

$\rho_{vp} = \rho_{vsat}(T_p)$. The similarity of heat and mass equations for similar boundary conditions is not applicable here to the extent that the boundary layers in temperature and humidity do not start at the same point (point 0 and point 1). There will therefore be 3 integral equations to solve: velocity, temperature, vapor density. The previous integral equations [2.22] and [2.23] will be slightly modified for curved surfaces to take into account a temperature condition valid at the wall. The buoyancy term that brings ΔT into play is modified in the equation of motion [2.22] which is thus written:

$$\begin{cases} a_1 \frac{\delta_+}{r_{a+}U_+} \frac{d}{dx_+}(r_{a+}U_+^2\delta_+) = Pr\frac{\delta_+^2}{U_+}\left[a_2 f_{Tp} sin\varphi + Ra^{-1/4}P_{hz}\right] - Prf'_u(0) \\ P_{hz} = cos\varphi[(a_3 + a_4)f_{Tp}\frac{d\delta_+}{dx_+} + a_3\delta_+\frac{df_{Tp}}{dx_+}] \end{cases} \quad [2.34]$$

P_{hz} is the horizontal contribution of buoyancy.

The energy conservation equation [2.23] becomes:

$$a_5 \frac{d}{dx_+}\left(r_{a+}U_+\delta_+ f_{Tp}\right) = \frac{r_{a+}}{\delta_+}(-f'_T(0))f_{Tp} \quad [2.35]$$

To these equations we should add a water vapor mass conservation equation similar to that of energy (with boundary layer hypotheses):

$$u\frac{\partial \rho_v}{\partial x} + v\frac{\partial \rho_v}{\partial y} = D_v\frac{\partial^2 \rho_v}{\partial y^2} \quad [2.36]$$

This equation shows that the order of magnitude of the evaporation boundary layer, denoted δ_c (often named "concentration" in the literature), is: $\delta_c^2 <\approx \frac{xD_v}{U}$ while the energy equation is: $\delta_T^2 <\approx \frac{x\alpha}{U}$. A comparison of the orders of magnitude of boundary layer thicknesses gives: $\frac{\delta_c^2}{\delta_T^2} \approx \frac{D_v}{\alpha} = \frac{Pr}{Sc}$ where the Schmidt number is $Sc = \upsilon/D_v$.

For the physical context under study we will take: $D_v = 2.6*10^{-5}$ m^2/s, $\upsilon = 1.58*10^{-5}$ m^2/s, therefore Sc = 0.61. The Prandtl and Schmidt numbers have the same order, so the boundary layer thicknesses are comparable. When these layers start at the same point, deducing the mass transfer coefficient from the heat transfer coefficient is sufficient. This is not the case here, so we should also solve an integral equation on the sphere, similar to the equation in temperature [2.35]:

$$a_5 \frac{d}{dx_+}\left(r_{a+}U_+\delta_{c+}f_{cp}\right) = \frac{r_{a+}}{\delta_{c+}}(-f'_c(0))f_{cp} \quad [2.37]$$

The profile of the boundary layer in humidity $\rho_{v+} = (\rho_v-\rho_{vo})/(\rho_{vsat}(T_p) - \rho_{vo}) = f_c(y/\delta_c)$ is considered identical to that in temperature (Figure 2.2): $\rho_{v+}(y/\delta_c) = T_+(y/\delta)$ which leads to $f'_c(0) = f'_T(0)$.

2.2.2. *Boundary layers for the horizontal disk*

The system of equations is somewhat simplified on the horizontal part where boundary layers are initiated. In effect, we consider $\varphi = 0$, f_{Tp} given [2.33], x the abscissa defined (Figure 2.5) by L = R and a radius vector degenerated in $r_a(x) = 2R+H-x$, or normed by D, $r_{a+}(x_+) = 1+H/D-x_+$. Taking into account $H = R\cos(\pi-\varphi_o)$, $\varphi_o = 110°$, $x_{2+} = 0.768$, $H/D = 0.171$.

The system [2.34], [235] becomes:

$$\begin{cases} a_1 \dfrac{\delta_+}{r_{a+}U_+}\dfrac{d}{dx_+}(r_{a+}U_+^2\delta_+) = PrRa^{-1/4}\dfrac{\delta_+^2}{U_+}P_{hz} - Prf'_u(0) \\ P_{hz} = (a_3 + a_4)f_{Tp}\dfrac{d\delta_+}{dx_+} + a_3\delta_+\dfrac{df_{Tp}}{dx_+} \end{cases} \quad [2.38]$$

$$a_5 \frac{d}{dx_+}\left(r_{a+}U_+\delta_+f_{Tp}\right) = \frac{r_{a+}}{\delta_+}(-f'_T(0))f_{Tp} \quad [2.39]$$

Similarly to the isolated sphere, the boundary layers in velocity and temperature start from a singular point. Let us look for an approximate solution in the vicinity of x = 0, starting from orders of magnitude of the terms of [2.38], [2.39]. If $r_{a+} = 1$; $f_{Tp} = \frac{x_+^2}{x_{2+}^2}$, we have:

$$inertia \approx viscosity \Rightarrow \frac{\delta_+}{U_+}\frac{d}{dx_+}(U_+^2\delta_+) \approx \frac{\delta_+^2 U_+}{x_+} \approx 1$$

$$buoyancy \approx viscosity \Rightarrow \frac{\delta_+^3}{U_+}x_+ \approx 1$$

The energy equation gives: $\frac{\delta_+^2 U_+}{x_+} \approx 1$. The compatibility of these relations leads to considering, similarly to the isolated sphere, $\delta = \delta_o$, $U_+ = cx_+$. Then, equations [2.38], [2.39] become:

$$2c\delta_o^2 = 2\frac{a_{10}}{x_{2+}^2}\frac{\delta_o^3}{c} - a_8 \; ; \; a_{10} = \frac{a_3}{a_1} PrRa^{-\frac{1}{4}} \; ; 3c\delta_o^2 = a_9$$

or:

$$\delta_0^2 c = \frac{a_9}{3} \; ; \; \frac{\delta_o^3}{c} = x_{2+}^2 \frac{a_8 + 2a_9/3}{2a_{10}}$$

Which allows us to calculate $\delta_o = 7.546$ and $c = 0.109$.

We can solve [2.38], [2.39] directly by correcting $d\delta_+/dx_+$ at each step or by decomposing the derivatives of U_+ and δ_+ as follows:

$$\left(U_+ - a_6 f_{Tp}\frac{\delta_+}{U_+}\right)\frac{d\delta_+}{dx_+} + 2\delta_+\frac{dU_+}{dx_+} = a_{10}f'_{Tp}\frac{\delta_+^2}{U_+} - \frac{U_+\delta_+}{r_{a+}}\frac{dr_{a+}}{dx_+} - \frac{a_8}{\delta_+} \quad [2.40]$$

$$U_+\frac{d\delta_+}{dx_+} + \delta_+\frac{dU_+}{dx_+} = -\frac{U_+\delta_+}{f_{Tp}}\frac{1}{r_{a+}}\frac{d(r_{a+}f_{Tp})}{dx_+} + \frac{a_9}{\delta_+} \quad [2.41]$$

2.2.3. *Boundary layers on curved surfaces*

The expression will be the same for a toroidal "coupling" surface and the upper half sphere. For convenience we will introduce the angle θ on the coupling, and it will play the role of φ (Figure 2.5). The previous dynamic equation is slightly modified:

$$\left(U_+ - a_6 f_{Tp}\frac{\delta_+}{U_+}\right)\frac{d\delta_+}{dx_+} + 2\delta_+\frac{dU_+}{dx_+} =$$
$$\frac{\delta_+}{U_+}\left(a_7 f_{Tp}\sin\theta + a_{10}f'_{Tp}\cos\theta.\delta_+\right) - \frac{U_+\delta_+}{r_{a+}}\frac{dr_{a+}}{dx_+} - \frac{a_8}{\delta_+} \quad [2.42]$$

The thermal equation is identical to [2.41].

On the coupling surface:

$$r_{a+} = \frac{1}{2} + \frac{H}{D}(1 - \sin\theta); \; x_+ = x_{1+} + \frac{H}{D}\theta \; ; \; x_{1+} = \frac{1}{2} \; ; 0 \le \theta \le \pi/2$$

$U_+(x_{1+})$, $\delta_+(x_{1+})$ are initial values resulting from solving the equations on the horizontal disk.

On the sphere:

$$r_{a+} = \frac{sin\varphi}{2}\ ;\ x_+ = x_{2+} + \frac{1}{2}\left(\frac{\pi}{2} - \varphi\right);\ x_{2+} = \frac{1}{2} + \frac{H}{D}\frac{\pi}{2}\ ;0 \leq \varphi \leq \frac{\pi}{2}$$

$U_+(x_{2+})$, $\delta_+(x_{2+})$ are initial values resulting from solving the equations on the coupling.

The water vapor transfer equation is similar to [2.39] with specific values $f_{cp}(x)$ and specific initial conditions. Having arranged the equation in view of its resolution, we obtain:

$$U_+ \frac{d\delta_{c+}^2}{dx_+} = -2\delta_{c+}^2 \left[\frac{dU_+}{dx_+} + U_+ \frac{1}{r_{a+}f_{cp}} \frac{d(r_{a+}f_{cp})}{dx_+}\right] + 2a_9 \qquad [2.43]$$

On the coupling: $\frac{1}{f_{cp}} \frac{d(r_{a+}f_{cp})}{dx_+} = \frac{2r_{a+}}{x_+ - x_{1+}} - cos\theta$.

On the sphere: $\frac{1}{f_{cp}} \frac{d(r_{a+}f_{cp})}{dx_+} = -cos\varphi$.

Let us note that equation [2.43] is a differential equation in δ_{c+}^2, $U_+(x_+)$ resulting from the resolution of previous equations [2.41], [2.42]. The initial condition $\delta_{c+}(x_{1+}) = 0$ corresponds to a singular point that we can comprehend by an order of magnitude reasoning. In effect, [2.43] gives in the vicinity of $x = x_1$, $\delta_{c+}^2 \approx x_+ - x_{1+}$. Let us consider $\delta_{c+}^2 = c(x_+ - x_{1+})$ in the equation and in the vicinity of x_{1+} :$c = \frac{2a_9}{5U_+}$, which enables the start of calculation.

Similarly to the isolated sphere, the numerical resolution of the system of three equations can be performed by a piece of software on the "market" of differential equations resolution, or by a program personally developed in a language (Fortran, basic, etc.) that one masters, or even more rapidly, on a spreadsheet. We then get functions $U_+(x_+)$, $\delta_+(x_+)$, $\delta_{c+}(x_+)$ starting from which we can deduce average Nusselt and Sherwood numbers.

Considering the definitions:

$$\begin{cases} \overline{Nu} = \frac{\bar{h}D}{\lambda} \; ; \; \overline{Sh} = \frac{\bar{k}D}{D_v} \; ; h(T_p - T_o) = -\lambda\left(\frac{\partial T}{\partial y}\right)_p = -\lambda\frac{T_p - T_o}{\delta}\left(\frac{\partial T_+}{\partial \eta}\right)_0 = 3\lambda\frac{T_p - T_o}{\delta} \\ \bar{h}\int_{\sigma\, courbes}(T_p - T_o)d\sigma = \int_{\sigma\, courbes} 3\lambda\frac{T_p - T_o}{\delta}d\sigma \; ; \; T_p - T_o = f_{Tp}(x)\Delta T \end{cases} \quad [2.44]$$

We get:

$$\overline{Nu}\int_{\sigma\, courbes} f_{Tp} r_{a+} dx_+ = 3Ra^{1/4}\int_{\sigma\, courbes} f_{Tp}\frac{r_{a+}}{\delta_+}dx_+ \quad [2.45]$$

where $\sigma_{courbes}$ is the followed coupling surface of the half sphere. Similarly:

$$\overline{Sh}\int_{\sigma\, courbes} f_{cp} r_{a+} dx_+ = 3Ra^{1/4}\int_{\sigma\, courbes} f_{cp}\frac{r_{a+}}{\delta_{c+}}dx_+ \quad [2.46]$$

Rapid calculation on a spreadsheet leads to: $\overline{Nu} = 0.48\, Ra^{1/4}$; $\overline{Sh} = 0.61 Ra^{1/4}$.

The difference in boundary conditions in temperature and humidity is reflected in the coefficient value as a factor of the Rayleigh number. The boundary layer in humidity starting in x_1 is less thick than the one in temperature that starts at x = 0, and is therefore more effective for the transfer.

2.3. Forced convection around a cylinder

This textbook case, well known to heat engineers for its many applications, is a good example for testing simplified methods for boundary layer integration or "integral" methods. In effect, for many problems of forced flow around various shapes, some experimental data are available (pressure distribution, detachment points, etc.) with the exception of transfer coefficients. Is it possible to rapidly derive good orders of magnitude of heat (and evaporation) transfers without resorting to specialized software in order to evaluate the relevance of results in the detachment configurations?

The integral method for the resolution of a flow around a cylinder, Karman–Pohlhausen type, is well known [SCH 60] but is limited to the calculation of laminar boundary layer until detachment point. Is it possible to extend the calculation of the dynamic boundary layer to the back region where wake originates? Is it possible to take into account the eventual transition of the boundary layer to turbulence (supercritical flow)?

The wake region behind the cylinder is the place where a path of counter-rotating vortices form, the so-called "Karman" vortices. The first two "lick" the wall of the back and give rise to a downwind boundary layer which goes up the wall, in opposition to the upwind boundary layer (Figure 2.6). The thermal boundary layers are linked to the dynamic boundary layers and the examination of local experimental data (friction, heat exchange) does not indicate any discontinuity at the passage of the detachment point. While the existence of dynamic and thermal boundary layers "upwind" is certain, from stagnation to detachment point, the nature of the back boundary layers "downwind" seems less obvious, due to the whirling nature of the external flow. In effect, the whirling path composed of big structures (vortex) intermingled with smaller ones is permanently evolving. We should consider that the inertia of 2 vortices that are close to the wall drives an external velocity at the boundary layer that we will comprehend by measuring pressure at the wall.

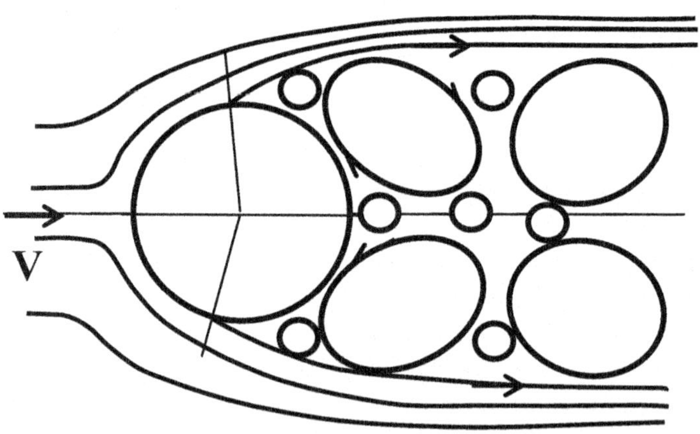

Figure 2.6. *Diagram of the flow around a cylinder*

2.3.1. *System of equations*

Let us suppose a cylinder with diameter D, radius R, temperature at the wall T_p, placed in a uniform flow of velocity V and temperature T_o. We will consider a flow of air, to be able to neglect the gravity effect. The equations of the boundary layer are obtained by simplifying the general equations of convection in cylindrical coordinates r, θ that we transform in coordinates $x = R\theta$ and $y = r-R$. For a steady flow with constant physical parameters (μ, λ, ρ, c_p), components of velocity vector $u = v_\theta$, $v = v_r$, pressure p, and temperature T, functions of x,y satisfy the following conservation equations for mass, momentum (over x) and energy. Let us note that here, unlike the case of natural convection, the gravity effect is neglected compared to pressure (low velocities of air).

$$\frac{\partial u}{\partial x} + \frac{\partial v}{\partial y} = 0 \qquad [2.47]$$

$$\rho\left(u\frac{\partial u}{\partial x} + v\frac{\partial u}{\partial y}\right) = -\frac{\partial p}{\partial x} + \mu\frac{\partial^2 u}{\partial y^2} \qquad [2.48]$$

$$\rho c_p\left(u\frac{\partial T}{\partial x} + v\frac{\partial T}{\partial y}\right) = \lambda\frac{\partial^2 T}{\partial y^2} \qquad [2.49]$$

The boundary layer approximations lead to setting p(x). Beyond a Reynolds number $R_e = VD/\upsilon$ of 10^5 the boundary layer can, starting from a certain point, become turbulent, which translates as a sudden increase in friction and transfer coefficients. The previous equations can be written in average values and, to keep it short, the quadratic terms can be regrouped as turbulent viscosity and turbulent thermal diffusivity. There is however a "laminar under-layer" in the immediate vicinity of the wall where the importance of turbulent effect decreases and at the wall the only playing factors are molecular viscosity and diffusivity. The "philosophy" of an integral method resides in considering only the integrated value, over thickness, of boundary layers and therefore proposing a possible profile form. Confronting result and experience can justify the relevance of the profile form and it is known that laminar and turbulent profiles are somewhat different.

To set a velocity profile $\frac{u}{U} = f_u\left(\frac{y}{\delta}\right)$ or a temperature profile $\frac{T-T_o}{T-T_p} = f_T\left(\frac{y}{\delta_T}\right)$ supposes well established definitions of U and T_o. For the "upwind" upstream part, we refer to a flow external to the boundary layer named non-viscous or with velocity potential which in this case is written

$\frac{U}{V} = 2sin\theta$ with, on a stream line, $p + \frac{1}{2}\rho U^2 = Cst$. For the downstream wake part, the situation is more complex. The two low-dissipative big vortices at the back, contrary to the small structures, impose a quasi-uniform field of velocity and pressure. In effect, the measured pressure coefficient $C_p(\theta) = \frac{p-p_o}{\frac{1}{2}\varrho V^2}$ varies very little along the cylinder surface of the detached region. At least in a first approximation we can admit there is on the outside of the rear boundary layer a (measured) velocity and pressure that satisfy $p + \frac{1}{2}\rho U^2 = Cst$ with $p(\theta) = p_o + \frac{1}{2}\varrho V^2 C_p(\theta)$. We have to admit as well that the cool air driven by these big vortices induces a temperature outside of the boundary layer which differs little from T_o. The setting up of integral equations will thus in this context concern both upwind and downwind boundary layers except that these layers develop in opposite directions on the cylinder surface starting from their respective origin $\theta = 0$ upwind and $\theta = \pi$ downwind.

2.3.2. *Integration of the equations of the dynamic boundary layer*

This approach is not very different from classical methodology, in the sense that integration is performed directly [2.47], [2.48] without going through the introduction of thicknesses of movement and momentum, and pressure is taken into consideration through the (measured) pressure coefficient.

Considering [2.47], the inertia term of [2.48] is written: $\frac{\partial u^2}{\partial x} + \frac{\partial uv}{\partial y}$ and the integration over y [2.48] then gives:

$$\frac{\partial}{\partial x}\int_0^\delta u^2 dy + [uv]_0^\delta = -\frac{1}{\rho}\frac{dp}{dx}\delta + \frac{\mu}{\rho}\left[\frac{\partial u}{\partial y}\right]_0^\delta \qquad [2.50]$$

Let us note that mathematical rigor requires the integration from 0 to L>δ before making L tend to δ.

Let us note u = v = 0 at y = 0;$[uv]_0^\delta = Uv(\delta), \tau_p = \mu\frac{\partial u}{\partial y}(y = 0)$ and let us *a priori* consider $\frac{\partial u}{\partial y}(y = \delta) = 0$.

The integration of [2.47] gives: $v(\delta) = -\int_0^\delta \frac{\partial u}{\partial x} dy$ which allows writing [2.50] as follows:

$$\frac{d}{dx}\int_0^\delta u^2 dy - U\frac{d}{dx}\int_0^\delta u dy = -\frac{\delta}{\rho}\frac{dp}{dx} - \frac{\tau_p}{\rho} \quad [2.51]$$

The profile of velocities is formally expressed $\frac{u(x,y)}{U(x)} = f(\eta, \beta(x))$; $\eta = \frac{y}{\delta(x)}$ where β(x) is a parameter that permits the distortion of the profile along the flow until detachment. This fact explains why integral methods are more difficult to implement in forced convection than in natural convection.

The following functions appear:

$$\int_0^1 f(\eta, \beta) d\eta = A_m(\beta); \int_0^1 f^2(\eta, \beta) d\eta = A_c(\beta) \quad [2.52]$$

The A_m function represents the average velocity normed by U in the thickness of the boundary layer and A_c is the normed average kinetic energy. Equation [2.51] becomes:

$$\frac{d}{dx}A_c U^2 \delta - U\frac{d}{dx}A_m U\delta = -\frac{\delta}{\rho}\frac{dp}{dx} - \frac{\tau_p}{\rho} \quad [2.53]$$

Velocity U(x) is linked to p(x) by:

$$p + \frac{1}{2}\rho U^2 = p_o + \frac{1}{2}\rho V^2; C_p = \frac{p - p_o}{\frac{1}{2}\rho V^2}; \frac{U}{V} = \sqrt{1 - C_p} \quad [2.54]$$

An additional relation is needed in order to get an equation with one unknown. The classical approach is to write the equation of motion [2.48] at the wall:

$$-\frac{dp}{dx} + \mu\left(\frac{d^2u}{dy^2}\right)_{y=0} = 0 \quad [2.55]$$

By introducing f(η, β):

$$\delta^2 \frac{dp}{dx} = \mu U\left(\frac{d^2 f}{d\eta^2}\right)_{y=0} \quad [2.56]$$

The first member of [2.53] can be developed as:

$$U^2\delta\frac{d}{dx}(A_c - A_m) + U^2\frac{d\delta}{dx}(A_c - A_m) + U\delta\frac{dU}{dx}(2A_c - A_m)$$

Equation [2.53] is then expressed with velocity U as a function of Cp by [2.54]:

$$(1-C_p)\left[\delta\frac{d(A_c-A_m)}{dx}+(A_c-A_m)\frac{d\delta}{dx}\right]-\frac{1}{2}\frac{dC_p}{dx}\delta(2A_c-A_m)= \\ -\frac{\delta}{2}\frac{dC_p}{dx}-\frac{\mu}{\rho}\frac{f'(0,\beta)}{\delta V}\sqrt{1-C_p} \qquad [2.57]$$

where $f'(0,\beta)=\frac{\partial f(\eta,\beta)}{\partial\eta}(0,\beta)$.

The weak point of the integral method applied to a flow under forced convection can be observed at this point. In effect, the method is an approximation based on an *a priori* set profile which avoids the exact determination of profile, which is numerically possible with a specialized software. The integration has to "erase" the differences between possible profiles. But associating a method of type [2.56] which relates to the second derivative at the wall is not in line with the "philosophy" of the method and it considerably narrows the range of profiles that can be proposed.

2.3.3. *Dimensionless integral equation*

Lengths are taken with respect to the diameter of the cylinder, which is correct for x but unusual for δ. We will see further on which would be the appropriate length of reference for δ. The dimensionless variables being $x_+=\frac{x}{D};\delta_+=\frac{\delta}{D}$, we have:

$$(1-C_p)\delta_+\frac{d(A_c-A_m)\delta_+}{dx_+}=\frac{\delta_+^2}{2}(2A_c-A_m-1)\frac{dC_p}{dx_+}-\frac{f'(0,\beta)}{R_e}\sqrt{1-C_p} \qquad [2.58]$$

This equation with two unknowns $\delta_+(x_+)$ and $\beta(x_+)$ is coupled to [2.56] which is written without dimensions and U expressed by C_p:

$$\delta_+^2=\frac{2f''(0,\beta)}{R_e}\frac{\sqrt{1-C_p}}{\frac{dC_p}{dx_+}} \qquad [2.59]$$

where: $f''(0,\beta)=\frac{\partial^2 f(\eta,\beta)}{\partial\eta^2}(0,\beta)$. Let us note $C_p(\theta)$ and $dx = Rd\theta$ so $dx_+ =$ $d\theta/2$.

For a given $C_p(\theta)$ and a deformable velocity profile with a $\beta(x)$ parameter, the system [2.58], [2.59] contains the solution $\delta_+(x_+)$, $\beta(x_+)$.

Several modifications of system [2.58], [2.59] can facilitate resolution. The boundary layer thickness δ_+ can appear only in squared form if we decompose the first member of [2.58]:

$$\frac{(1-C_p)}{2}(A_c-A_m)\frac{d\delta_+^2}{dx_+}=\delta_+^2[\frac{1}{2}(2A_c-A_m-1)\frac{dC_p}{dx_+}$$
$$-(1-C_p)\frac{d(A_c-A_m)}{dx_+}]-\frac{f'(0)}{R_e}\sqrt{1-C_p} \qquad [2.60]$$

Let us divide the terms of this last equation by $\sqrt{1-C_p}$ and multiply by R_e so that a new variable $X = R_e\delta_+^2$ appears. This shows that a good length of reference for δ would be: $D/R_e^{1/2}$. We have:

$$\frac{\sqrt{1-C_p}}{2}(A_c-A_m)\frac{dX}{dx_+}=X[\frac{(2A_c-A_m-1)}{2\sqrt{1-C_p}}\frac{dC_p}{dx_+}-\sqrt{1-C_p}\frac{d(A_c-A_m)}{dx_+}]-f'(0,\beta) \qquad [2.61]$$

and [2.59] becomes:

$$X=2f''(0,\beta)\frac{\sqrt{1-C_p}}{\frac{dC_p}{dx_+}} \qquad [2.62]$$

An equation with one unknown $\beta(x)$ is formally derived. Let us note that this system [2.61], [2.62] is also valid in a turbulent context, since the integration over the boundary layer only retains the friction to the wall expressed similarly in laminar and turbulent context, and "turbulent friction" is only taken into account by the velocity profile *a priori* set.

To derive a more compact form of [2.61], [2.62] let us consider:

$$a_1(x_+)=\frac{\sqrt{1-C_p}}{2};\, a_2(x_+)=\frac{1}{2\sqrt{1-C_p}}\frac{dC_p}{dx_+};\, b_1(\beta)=$$
$$A_c-A_m;\, b_2(\beta)=2A_c-A_m-1 \qquad [2.63]$$

The elimination of $X(x_+, \beta)$ between [2.61] and [2.62] leads to:

$$\frac{a_1}{a_2}\left(b_1\frac{df''(0,\beta)}{d\beta}+2b'_1f''(0,\beta)\right)\frac{d\beta}{dx_+}=\left(a_1\frac{a'_2}{a_2^2}b_1+b_2\right)f''(0,\beta)-f'(0,\beta) \qquad [2.64]$$

2.3.4. *Resolution of the upwind dynamic boundary layer*

Let us first consider the problem of the "upwind" dynamic boundary layer before extrapolating to the "downwind" layer. The "upwind" starting point, origin of the boundary layer at x = 0, is a singular point defined here by $C_p = 1$. In the vicinity of the stagnation point the solution to non-viscous flow "sticks" perfectly:

$$C_p = 1\text{-}4\sin^2\theta;\ a_1 = \sin\theta;\ \frac{dC_p}{dx_+} = -16\sin\theta\cos\theta;\ a_2 = -4\cos\theta;\ \frac{da_2}{dx_+} = 8sin\theta.$$

Equation [2.64] will give the conditions at the origin θ = 0 (or x = 0) for β and its derivative. In effect, for θ = 0 this equation is degenerate. The first member is zero and the second member reduces to:

$$F(\beta) = b_2 f''(0,\beta) - f'(0,\beta) = 0 \qquad [2.65]$$

This sets a fixed initial value for β, but does not impose any condition on the function f(η, β).

The derivative of β at the origin is undetermined, but the division of the two members of [2.64] by sinθ leads to applying l'Hôpital's rule to the ratio $\frac{F(\beta)}{sin\theta} = \frac{F\prime(\beta)\frac{d\beta}{d\theta}}{cos\theta}$. A little shaping shows that the derivative at the origin is zero.

Carrying on the calculation requires a datum of the function $C_p(x)$ and a type of velocity profile $\frac{u}{U} = f(n,\beta)$. Let us consider a flow around a cylinder for a Reynolds number $10^4<R_e<10^5$ so that the boundary layer stays laminar before detachment (Figure 2.7) [SAD 87].

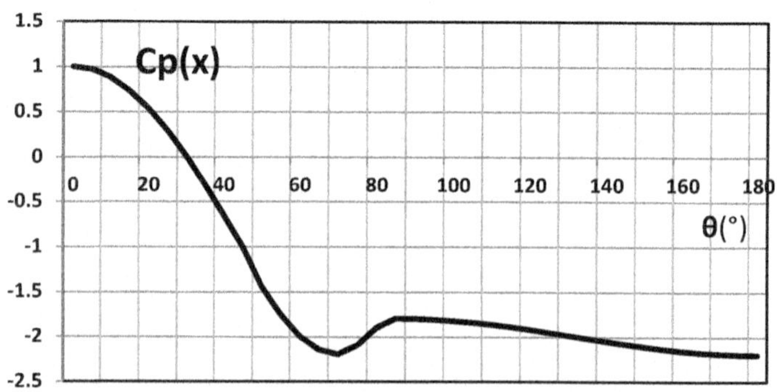

Figure 2.7. *Pressure coefficient around a cylinder*

It is convenient to translate this curve into parts of equations that preserve the continuity of the first derivative but not necessarily that of the second derivative. Some parts of a sinusoid fit well on the intervals [0, 45°]; [45°, 70°]; [70°, 85°]; [85°, 180°]. The point of detachment is at 85°.

$0<\theta<45° \rightarrow C_p = 1\text{-}4\sin^2\theta$; $C_p(45°) = -1$; $C_p'(45°) = -4$

$C_p(70°) = -2.2$; $C_p'(70°) = 0$; $C_p(85°) = -1.8$; $C_p'(85°) = 0$; $C_p(180°) = -2.2$; $C_p'(180°) = 0$.

Let us now search for a velocity profile $f(\eta, \beta)$ that satisfies $f(0, \beta) = 0$; $f(1, \beta) = 1$; $\frac{df}{d\eta}(1,\beta) = 0$; and naturally $0 \leq f(\eta,\beta) \leq 1$. A polynomial function is the usual choice but the position of parameter β in the polynomial is also a matter of choice and it would make better physical sense to have β equal to the derivative at the wall: $\frac{df}{d\eta}(0,\beta) = \beta$. The point of detachment then corresponds to a zero value for beta. To each degree of the polynomial there corresponds a higher limit of β starting from which the function exceeds one unit in the vicinity of $\eta = 1$. The 4 degree profile retained here ($\beta < 4$) requires an additional boundary condition which will be: $\frac{\partial^2 f}{\partial \eta^2}(1,\beta) = 0$.

Satisfying the boundary conditions set at the wall and at the border of the boundary layer leads to:

$$f(\eta,\beta) = \beta.\eta + 3(2-\beta)\eta^2 + (3\beta - 8)\eta^3 + (3-\beta)\eta^4 \qquad [2.66]$$

Figure 2.8 shows the evolution of the normed velocity profile, to be determined, from the stagnation point ($\theta = 0$, a) to the detachment point ($\theta = 85°$, b). An informed reader notices that these profiles are not adequate to turbulent profiles, where friction to the wall is much higher. We should therefore "cut" this profile into pieces (laminar under-layer, turbulent under-layer, external layer) assigning equation forms to each of them.

It follows:

$$A_m = \int_0^1 f d\eta = \frac{3}{5} + \frac{\beta}{20} \text{ ; } A_c = \int_0^1 f^2 d\eta = \frac{17}{35} + \frac{34}{840}\beta + \frac{1}{252}\beta^2$$

$$b_1 = c_1 + c_2\beta + c_3\beta^2;\ c_1 = -\frac{4}{35};\ c_2 = -\frac{1}{105};\ c_3 = \frac{1}{252}$$

$$b_2 = c_4 + c_5\beta + 2c_3\beta^2\ ;\ c_4 = -\frac{22}{35}\ ;\ c_5 = \frac{13}{420}$$

$$f''(0,\beta) = 6(2-\beta)$$

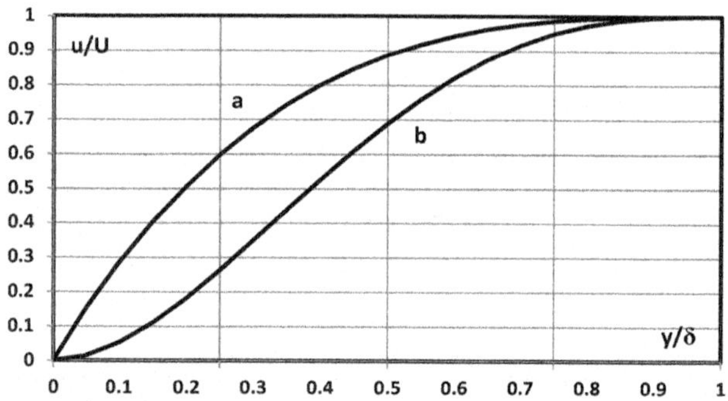

Figure 2.8. *Velocity profiles for β = 3.17 a) and β = 0 b)*

The solution to equation [2.65], of third degree in β, gives the initial value $\beta_o = 3.17539$. The 2 other roots are not acceptable, one is negative and the other one allows for values of f that are higher than 1.

An approximate solution in the vicinity of θ = 0 is deduced if we consider the importance of various terms of equation [2.64]. In effect the second member of this equation is a function of β if we can neglect the term in θ: $\left|a_1 \frac{a'_2}{a_2^2} b_1\right| \ll |b_2|$.

The term in θ is written: $a_1 = \sin\theta$; $a_2 = -4\cos\theta$; $a'_2 = 8\sin\theta$; $a_1 a'_2/a_2^2 = tg^2\theta/2$ which in the vicinity of zero is developed as: $\frac{tg^2(\theta)}{2} = \frac{1}{2}\theta^2 + \frac{1}{3}\theta^4 + \cdots$

The terms in β are developed in the vicinity of β_o:

$b_1(\beta) = b_1(\beta_o) + \varepsilon_\beta b'_1(\beta_o) + \cdots;\ b_2(\beta) = b_2(\beta_o) + \varepsilon_\beta b'_2(\beta_o) + \cdots;\ \varepsilon_\beta = \beta - \beta_o.b_1(\beta_o) = -0.1045;\ b'_1(\beta_o) = 0.0156; b_2(\beta_o) = -0.4502;\ b'_2(\beta_o) = 0.0813$

The first approximation of $\left|a_1 \frac{a'_2}{a_2^2} b_1\right| \ll |b_2|$ is written: $\left|\frac{\theta^2}{2} b_1(\beta_o)\right| \ll |b_2(\beta_o)|$ and taking into account the numerical values it is confirmed with a precision of about 3% up to an angle θ of 30° (π/6).

Equation [2.64] is then reduced and can be solved by parts:

$$\frac{b_1 \frac{df''(0,\beta)}{d\beta} + 2b'_1 f''(0,\beta)}{b_2 f''(0,\beta) - f'(0,\beta)} d\beta = \frac{a_2}{a_1} dx_+ \qquad [2.67]$$

The ratio in the first member of [2.67] is written with the function f chosen: $\frac{-6b_1 + 12(2-\beta)b'_1}{6(2-\beta)b_2 - \beta}$

The functions of β at the numerator F_{nu} and at the denominator $F_{dé}$ are to be developed in the vicinity of βo:

$$F_{nu} = 24c_2 - 6c_1 - 6(3c_2 - 8c_3)\beta - 30c_3\beta^2$$

$$F_{dé} = 12c_4 + 6\left(2c_5 - c_4 - \frac{1}{6}\right)\beta + 6(4c_3 - c_5)\beta^2 - 12c_3\beta^3$$

$$F(\beta) = F(\beta_o) + (\beta - \beta_o)F'(\beta_o) + \cdots$$

$$F_{nu}(\beta) = 0.4059 - 0.3941(\beta - \beta_o) + \cdots$$

$$F_{dé}(\beta) = 1.1278(\beta - \beta_o) + \cdots$$

Equation [2.67] becomes at first order in the vicinity of θ = 0, β = β_o ($dx_+ = d\theta/2$):

$$0{,}3599 \frac{d\beta}{\beta - \beta_o} = -2cotg\theta.d\theta$$

which is integrated in: $Ln(\beta - \beta_o) = -5.5562Ln(sin\theta) + cst$, which gives:

$$\beta - \beta_o = -sin^n\theta; n = 5.5562 \qquad [2.68]$$

Let us note $0 \leq \beta \leq \beta_o$.

The numerical resolution of [2.64] is to be performed by parts starting from the equations of parts of sinusoids by intervals. For the interval θ = 0 to

$\theta = 45°$ the numerical solution is very close to the approximate solution. For the interval $\theta = 45°$, $\theta = 70°$ the change of equation $C_p(\theta)$ does not assure the continuity of $C''_p(\theta)$, but does not disturb the numerical calculation. To the point $\theta = 70°$ there corresponds $C'_p = 0$, $a_2 = 0$, $f''(\beta) = 0$ by [2.62] so $\beta = 2$.

The resolution on the interval $\theta = 70°$, $\theta = 85°$ raises problems. In effect for $\theta = 85°$, we set $C'_p = 0$ or $a_2 = 0$ and $\beta = 0$ corresponding to the point of detachment. Function $f''(0, \beta)$ has to be zero for $\theta = 70°$ and $85°$ to satisfy equation [2.62]: $f''(0, \beta) = Xa_2(x)$. At the point of detachment the calculation may give an infinite boundary layer thickness, compatible with the set model, but the "velocity profile" function [2.66] cannot satisfy a regular evolution of β from 2 to 0 over this part of the contour. The search for a more complex profile is a delicate operation that complexifies and ruins the integral method. In effect, this method is simple to the extent that an *a priori* fixed profile gives a solution which is "not too distant" from the exact solution. The sign-changing gradient of pressure associated with the point of detachment no longer corresponds to a simple profile. The calculation of the boundary layer can be then carried on by another method. We will give up trying to satisfy the surface condition on the second derivative [2.62], which raises problems, but we will set a decreasing function for β from $\beta = 2$ (70°) to $\beta = 0$ (85°), such as:

$$\beta = \frac{2}{d}(x_+(85°) - x_+); d = x_+(85°) - x_+(70°).$$

Equation [2.61] then becomes a differential equation with one unknown under the form:

$$a_1 b_1 \frac{dX}{dx_+} = X\left(a_2 b_2 - 2a_1 b'_1 \frac{d\beta}{dx_+}\right) - \beta \qquad [2.69]$$

for which numerical resolution is easy, even on a spreadsheet.

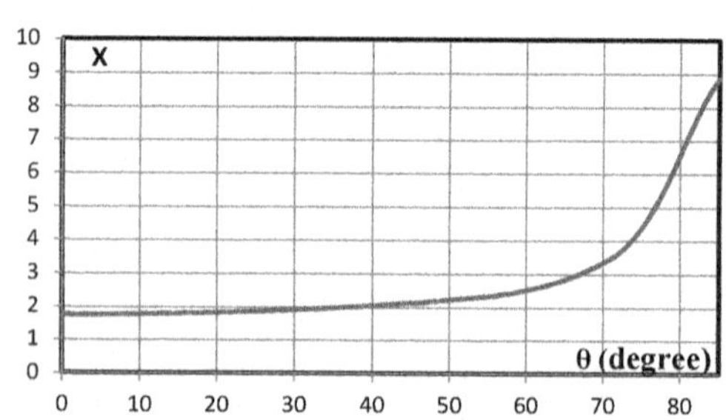

Figure 2.9. *Thickness of the "upwind" normed laminar boundary layer*

2.3.5. *Resolution of the downwind dynamic boundary layer*

Determining boundary layers on the back of a cylinder under wake can be conducted by using the same methodology. In effect, detachment generates large whirling structures that give rise to particular boundary layers in velocity and temperature, whose external border is imposed by a field of velocities that is comprehended only through the pressure at the wall. We should hypothesize that the unsteady fluctuations of velocity and pressure at the back have little influence on the previously introduced profile of velocities, which is not exact for a turbulent boundary layer. Let us note that, by nature, this dynamic boundary layer is influenced by wake vortices, unsteady flow, but not by the upstream boundary layer detached, be it laminar or turbulent.

This being set, profile [2.66] is applied to equations [2.61], [2.62] except that the evolution of the dynamic boundary layer is calculated from $\theta = 180°$ to $\theta = 85°$. The discretization will be made "backward" from x_+ to $x_+ - \Delta x_+$ starting from $x_+ = \pi$. The points of departure and of arrival are still characterized by $C'p = 0$ or $a_2 = 0$. Equation [2.62] imposes $f''(0, \beta) = 0$ in these 2 points, or $\beta = 2$ at the point of departure ($\theta = 180°$) and $\beta = 0$ at the point of detachment ($\theta = 85°$) but between these 2 points it is still too difficult to satisfy. As previously, we will set a linear variation of β in order to numerically solve equation [2.69] and the result is given in Figure 2.10.

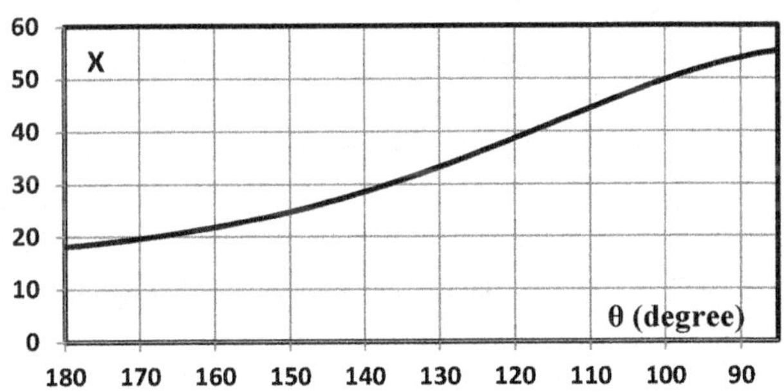

Figure 2.10. *Thickness of the "downwind" (normed) dynamic boundary layer*

The thickness of the "downwind" boundary layer is much higher than the one "facing the wind". A small calculation of the order of magnitude will indicate the main parameter coming into play. The boundary layer in forced convection can be described in terms of order of magnitude as the place where the pressure-linked forces compensate the viscosity forces which is (equation [2.48]):

$$\frac{\partial p}{\partial x} \approx \mu \frac{\partial^2 u}{\partial y^2} \approx \mu \frac{U}{\delta^2}$$

Taking into account that: $C_p = \frac{p-p_o}{1/2\rho V^2}; U = V\sqrt{1 - C_p}$, the previous relation becomes:

$$\delta^2 \approx 2\frac{\mu}{\rho}\frac{\sqrt{1-C_p}}{V\frac{dC_p}{dx}} \rightarrow \delta_+^2 \frac{VD}{\upsilon} \approx 2\frac{\sqrt{1-C_p}}{\frac{dC_p}{dx_+}}$$

or

$$X(x_+) = \delta_+^2 \frac{VD}{\upsilon} \approx \frac{1}{a_{2(x+)}}$$

This relation can be associated with [2.62] or $X = f'(0, \beta)/a_2$.

The back region is characterized by a small variation of C_p and thus by a value of a_2 approximately ten times weaker than in the upwind region.

2.3.6. *Resolution of the thermal boundary layer*

Having derived the dynamic boundary layers, the determination of the thermal boundary layer is quite easy. We should note only one thermal boundary layer, as its continuity is *a priori* assured at the point of detachment. The thickness of the thermal boundary layer for a Prandtl number close to the unit has *a priori* an order of magnitude similar to the dynamic boundary layer and this will be clarified when writing the equation of energy integrated over the boundary layer. This equation applicable to the boundary layer, with the approximations made, is written:

$$u\frac{\partial T}{\partial x} + v\frac{\partial T}{\partial y} = \alpha \frac{\partial^2 T}{\partial y^2} \qquad [2.70]$$

The boundary conditions are: $T(x, 0) = T_p$; $T(x, y = \delta_T) = T_o$; $T_p - T_o = \Delta T$ and taking into account the equation of mass conservation [2.47] the previous equation can be integrated over the thermal boundary layer in:

$$\frac{d}{dx}\int_0^{\delta_T} u(T - T_o)dy = -\alpha \left[\frac{\partial(T-T_o)}{\partial y}\right]_{y=0} \qquad [2.71]$$

The dimensionless temperature profile introduced needs to satisfy the previous boundary conditions and 2 additional conditions: second derivative is zero at the wall, derivative is zero at the external border. We get:

$$\eta_T = \frac{y}{\delta_T};\ f_T = \frac{T-T_0}{\Delta T};\ f_T(\eta_T) = 1 + \frac{\eta_T}{2}(\eta_T^2 - 3) \quad [2.72]$$

Or if we replace u and $T\text{-}T_o$ by f and f_T in [2.71]:

$$\frac{d}{dx}U\delta_T \int_0^1 f(\eta_T, \beta).f_T(\eta_T)d\eta_T = -\frac{\alpha}{\delta_T}\left(\frac{df_T}{d\eta_T}\right)_{\eta_T=0} \quad [2.73]$$

The integration of [2.73] raises the problem of the value of $\delta(\delta_T)$, still unsolved at this stage. Taking into account the same order of magnitude of δ and δ_T, let us hypothesize that the integral of [2.73] is to a little extent influenced by their difference. A correction loop can be established starting from function $\delta_T(x)$ which follows there from. Then the calculation of the integral does not raise any problem and [2.73] becomes:

$$\delta_T \frac{d}{dx}\left((3\beta + 13)U\delta_T\right) = 168\alpha \quad [2.74]$$

The dimensionless formulation is derived similarly to [2.61] by expressing $U(C_p)$, x_+, δ_{T+} and we get:

$$a_1(3\beta + 13)\frac{dX_T}{dx_+} + X_T\frac{d}{dx_+}(2a_1(3\beta + 13)) = \frac{168}{Pr} \quad [2.75]$$

where: $X_T = R_e\delta_{T+}^2$; $a_1 = \frac{\sqrt{1-C_p}}{2}$.

This type of differential equation can be integrated and the general solution can be written:

$$X_T = \frac{1}{(a_1(3\beta+13))^2}\left[\frac{168}{Pr}\int a_1(3\beta + 13)dx_+ + Cst\right] \quad [2.76]$$

The thermal boundary layer on the "upwind" part of the cylinder starts at $\theta = 0$ with $a_1 = 0$. The solution of $a_1 = \sin\theta$ in the vicinity of zero allows us to remove the root uncertainty and [2.76] becomes:

$$X_T(x_+) = \frac{1}{(a_1(3\beta+13))^2}\left[\frac{168}{Pr}\int_0^{x_+} a_1(3\beta + 13)dx_+\right] \quad [2.77]$$

where $X_T(0) = \frac{84}{Pr}\frac{1}{3\beta_o+13}$.

On the back of the cylinder ($\theta = 85°$ à $\theta = 180°$), following x_+ as it increases, starting from the point of detachment x_{+d}, so u<0, the equation of the boundary layer is:

$$X_T(x_+) = X_T(x_{+d}) - \frac{168}{Pr}\frac{1}{(a_1(3\beta+13))^2}\left[\int_{x_{+d}}^{x_+} a_1(3\beta + 13)dx_+\right] \quad [2.78]$$

Figure 2.11 shows this normed function of the thickness of the thermal boundary layer.

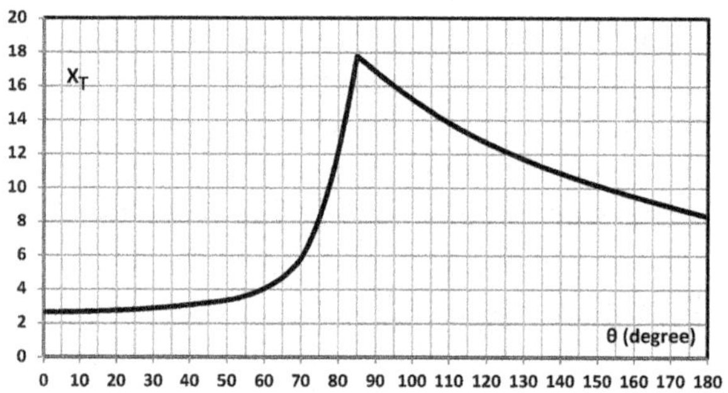

Figure 2.11. *Thermal boundary layer around the cylinder*

We note that on the "upwind" part X_T is hardly bigger than X while on the rear side of the back "downwind", X is clearly bigger than X_T. This is due to the factor $a_2(x_+)$ which magnifies X in the wake region, as seen previously, while the continuity of the thermal boundary layer adjusts its decreasing according to [2.78]. It would then make more "physical sense" to speak of increase starting from $\theta = 180°$ up to $\theta = 85°$ accompanying the dynamic boundary layer. We verify that the hypothesis $\eta_T = \eta$ in the integral of [2.73] is perfectly justified. In effect the iteration starting from values derived for X and X_T, thus δ and δ_T, modifies the results only to a little extent (<1%).

From X_T we deduce the local Nusselt number. From the following formulations:

$$Nu = \frac{hD}{\lambda};\, h\Delta T = -\lambda\left(\frac{\partial T}{\partial y}\right)_{y=0} = \frac{3}{2}\lambda\frac{\Delta T}{\delta_T};\; X_T = R_e\delta_{T+}^2$$

We get:

$$Nu\, R_e^{-1/2} = \frac{3}{2} X_T^{-1/2} \qquad [2.79]$$

Figure 2.12 shows the evolution of this function around the cylinder. The aspect is the one derived from measurements [SAD 87], which is considerable given the numerous hypotheses made. The correlations given in the literature (laminar/turbulent) are often in terms of $R_e^{-1/2} Pr^{-1/3}$. If we clarify here the Prandtl number from [2.77], [2.78] we should get $Pr^{-1/2}$ instead of $Pr^{-1/3}$. We should note that $a_1(\theta) = \frac{\sqrt{1-C_p}}{2}$ is the main parameter adjusting the curve.

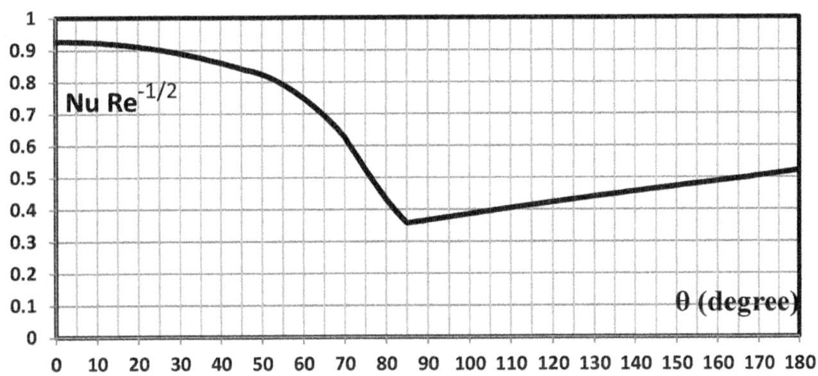

Figure 2.12. *The evolution of Nusselt number around the cylinder*

The experimental curves given in the literature [SAD 87] for the retained range of Reynolds numbers (10^4-10^5) give slightly higher values. We should note that in the vicinity of the stagnation point (0-30°), on a part of the still laminar boundary, the measured function $NuRe^{-1/2}$ varies with the Reynolds number (10^3, 10^6), which means that the profiles for velocity and temperature may be more complex than the ones set here.

3

Human Thermal Models

Adaptation of man to his climatic environment is a matter of concern with many facets: social (habitat), industrial (work or leisure environments, clothing) and… in fine, scientific. Extreme examples are numerous: hot work (iron and steel industry), expeditions to cold regions (mountains and poles), confined cockpits (submarine and space shuttle). The design of work environments (offices and factories) or transport vehicles (automobiles and trains) requires detailed studies of thermal comfort.

Insofar as the human body adapts to a certain extent to the changes in the environment (temperature, humidity, sunlight, etc.) through adequate physiological reactions (shiver, transpiration and respiration) and adapted behaviors (clothing), human thermal models are at the crossroads of physiology (internal regulation) and heat engineering (external factors and clothing). As for any physical model, the physical and physiological analysis should be conceived according to the proposed objective. It is *a priori* simpler to study the thermal comfort of a passenger traveling by HST than by space shuttle. We will examine the case of a baby who runs the risk of death due to excessive covering when put to sleep.

3.1. The Fanger model: from climatic chamber to standard [FAN 70]

In the 1970s a team of Danish physiologists led by Professor Fanger used the climatic chamber, a small room with controlled physical parameters, to test the notion of thermal comfort on human subjects at rest. A physical model for the thermal exchanges between man and room environment was established and a scale of thermal sensations was proposed, depending on physical parameters. These studies have been validated and resumed in the standard [ISO 7730].

We will look into this model, whose objective is to characterize an environment that is "fairly close" to comfort. We will later on clarify the notion of comfort as "statistical" satisfaction that results from numerous tests conducted in the climatic chamber with voluntary subjects.

3.1.1. *Environment and human body physical parameters*

Bearing in mind the climatic chamber, the physical parameters to be controlled are first those characteristic to the air: temperature, humidity, velocity. These parameters are uniform throughout the volume therefore thermal stratifications typical of common premises are excluded. We then follow the temperatures of the walls, with their emission properties, which allow for the evaluation of radiation exchange with an individual at rest in the room.

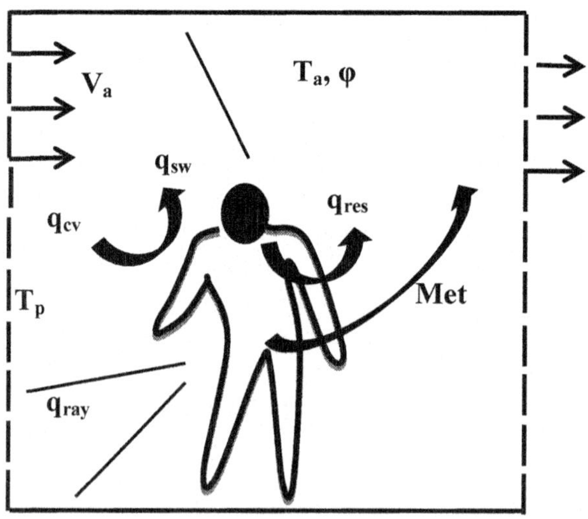

Figure 3.1. *Thermal exchanges of an individual in a climatic chamber*

Man is a particular "thermal machine" that produces heat (through metabolism), as the result of chemical oxidation at cell level, and is fitted with thermal sensors that activate regulation mechanisms (sweating, vasodilatation, etc.). Any effort undertaken (work) corresponds to an increase in metabolism, though a part of it goes back to the external environment. This metabolic heat is dissipated through skin and eventually across clothing, by convection, radiation and evaporation (transpiration). Finally, respiration is a cooling source that will need to be considered.

This model expresses these various heat fluxes for an individual as a whole, and it is a great simplification of a heterogeneous ensemble. A look at the notion of skin temperature serves well as illustration. While it is a well known fact that a healthy individual's temperature is around 37°C, this so-called internal temperature covers a certain range of skin temperatures that vary with the external one. In an environment at 20°C, skin temperature may display 34°C on the head, 33°C on the chest, 30°C on the hands, 25°C on the legs… Complex hygro-thermal exchanges across clothing will similarly be globalized in a simple formulation. Human body is here considered a "potato"-type homogenous ensemble with an internal heat source and a crossing over respiratory flow. Let us detail these various fluxes. As a matter of respect for the physiologists' practice, in this chapter fluxes will be noted without the dot accent (q instead of $\dot{q}$).

3.1.1.1. *Internal heat production or metabolism, denoted Met*

It is common practice to denote the heat flux in W/m^2 of skin, knowing that the skin surface area of a typical individual is 1.8 m^2. Here are some data for an adult metabolic production as a function of various activities (Table 3.1). For convenience reasons we are introducing a unit of measurement, called met, which is 58 W/m^2 and corresponds to the production of a sitting person at rest.

Activity	Lying down/at rest	Sitting/at rest	Office work sitting	Light work standing	Domestic work standing	Hard work standing
met	0.8	1	1.2	1.6	2	3
Met (W/m^2)	47	58	70	93	117	175

Table 3.1. *Metabolism*

Let us give some examples: manual dishwashing: 2.5 met; digging the garden: 6.5 met; running at 15 km/h: 9.5 met.

In the case of exercise that involves energy transfer (bicycle ergo meter, for example) some energy can be transmitted and needs to be recorded. Let us note that human machine has low efficiency (below 20%), which means that a significant part of the effort, here denoted W (W/m^2), is converted into heat.

For readers' convenience, when writing fluxes we can introduce notations that relate to English words (clothes, skin, etc.), similarly to the notations used in international standards. These fluxes are thus expressed in Watt per m^2 of skin.

3.1.1.2. *Convective losses*

Convective losses from the external surface of clothes at temperature T_{cl} considered uniform around the "potato" volume can be expressed as:

$$q_{cv} = f_{cl} h_{cv} (T_{cl} - T_a) \qquad [3.1]$$

where h_{cv} is a transfer coefficient (natural or forced convection) to be evaluated. The Fanger formulation (one among many others) is the following:

$$h_{cv}(W/m^2K) = max[2.38(T_{cl} - T_a);\ 12.1\sqrt{V_a}] \qquad [3.2]$$

The maximum of the two values is a choice between natural convection and forced convection. Parameter f_{cl} is a coefficient equal to the ratio between the external surface of the cloth under convection and skin surface area. This coefficient will be further on given as a function of clothing thermal resistance. It should be recalled that all fluxes are given per unit area (m^2) of skin surface. V_a is the (relative) velocity of incident air on the individual.

3.1.1.3. *Radiation losses*

In a framework of highly simplified physical phenomena, losses due to radiation from the external surface of clothing can be expressed as:

$$q_{ray} = f_{eff} f_{cl} \varepsilon \sigma (T_{cl}^4 - T_p^4) \qquad [3.3]$$

where f_{eff} is a coefficient that takes into account a surface reduction, given that not all the clothing surface has exchanges with the premise walls. This factor is experimentally determined, for example $f_{eff} = 0.696$ for a sitting person and 0.725 for a standing person.

ε is the emittance of the clothing surface, approaching one unit for long wavelengths of infrared radiation (0.97 in general) while σ is Stefan–Boltzmann constant $\sigma = 5.67.10^{-8}$ W/m²K⁴. All temperatures are expressed in Kelvin. T_p is the "average" temperature of the walls. It is clear that the exact problem is more complex and should take into account solid angles (shape factors) between various surfaces and their emissivity. For a rapid still sufficiently precise approach, we will use for this model the relations between "black" surfaces as is the case of a subject surrounded by a reduced number i of walls at temperature T_i and a shape factor F_{pi} with the subject ($\sum_i F_{pi} = 1$).

$$T_p^4 = \sum_i F_{pi} T_i^4 \qquad [3.4]$$

In inhabited premises this radiation factor greatly influences comfort. Walls can be classified as cold walls (walls, bay windows, "chilled" cooling beams, etc.) and hot walls (heating appliances, heated floors, etc.). Equation [1.3] is usually linearized as:

$$q_{ray} = f_{cl} h_{ray}(T_{cl} - T_p);\ h_{ray} = \varepsilon\sigma f_{eff}(T_{cl} + T_p)(T_{cl}^2 + T_p^2) \qquad [3.5]$$

EXAMPLES 3.1.– let us consider a dressed subject with $\varepsilon = 0.97$; $f_{eff} = 0.7$. We will indicate several temperatures in °C that will be transformed in Kelvin for the calculation.

$T_{cl} = 20$°C, $T_p = 10$°C→$h_{ray} = 3.7$ W/m²K; $T_{cl} = 25$°C, $T_p = 20$°C→$h_{ray} =$ 4 W/m²K;

$T_{cl} = 10$°C, $T_p = 0$°C→$h_{ray} = 3.3$ W/m²K; $T_{cl} = 10$°C, $T_p = 10$°C→$h_{ray} =$ 3.1 W/m²K;

For a naked subject: $\varepsilon = 0.97$; $f_{eff} = 0,7$: $T_{cl} = 34$°C, $T_p = 20$°C→$h_{ray} =$ 4.1 W/m²K.

3.1.1.4. *Heat flux conducted through clothing*

The fluxes convected and radiated by the clothing surface are transmitted through conduction across clothing.

$$q_{conduc} = (T_{sk} - T_{cl})/R_{th} \qquad [3.6]$$

where R_{th} is the clothing thermal resistance, measurable on the test bench. Comprehensive data are available for textile layer thermal resistances measured on samples, but what is needed here is a global thermal resistance for a highly heterogeneous ensemble with many "parasitic" phenomena such as air circulation between layers starting from the openings. The proper method is thus to perform global measurements on a heated mannequin (fixed skin temperature) placed in the climatic chamber. We thus define clothing "classes" that allow for *a priori* evaluation of clothing. A unit of resistance is introduced: the clo (I_{cl}) such that $R_{th} = 0.155\ I_{cl}$. Table 3.2 gives some examples of clothing and their corresponding value in clo.

Clothing Type	I_{cl} (clo)	R_{th} (m^2K/W)
Light summer beachwear	0.1	0.047
Summer daywear	0.5	0.078
Mid-season indoor work wear	0.8	0.124
Winter work wear with jacket	1	0.155
Winter wear	1.5	0.233

Table 3.2. *Clothing classes*

The clothing type determines the increase of the surface of exchange with the environment, that is the coefficient f_{cl}. We have:

$$I_{cl}<0.5\ \text{clo} \rightarrow f_{cl} = 1+0.2\ I_{cl};\ I_{cl}>0.5\ \text{clo} \rightarrow f_{cl} = 1+0.1\ I_{cl} \quad [3.7]$$

The flux conducted across the clothes is the sum of fluxes received by convection and radiation:

$$q_{conduct} = q_{cv} + q_{ray} \quad [3.8]$$

From [3.1], [3.5], [3.8] we derive:

$$q_{conduct} = f_{cl}F_{cl}(h_{cv} + h_{ray})(T_{sk} - T_{op}) \quad [3.9]$$

F_{cl} is the (dimensionless) reduction factor due to clothing:

$$F_{cl} = \left[1 + 0.155 I_{cl} f_{cl}(h_{cv} + h_{ray})\right]^{-1} \quad [3.10]$$

T_{op} is the operative temperature:

$$T_{op} = \frac{h_{ray}T_p + h_{cv}T_a}{h_{ray} + h_{cv}} \quad [3.11]$$

Thus we have simply written a sum of thermal resistances of clothing and of the external boundary layer around clothing.

3.1.1.5. *Evaporative flux*

This is the model's difficult point. In effect, the flux evaporated from humid skin is an important regulating mechanism in certain contexts (sports practice, hard work, strong heat, etc.) and the clothing protection is more or less effective as a water vapor filter. According to the original Fanger model, upon which standard ISO 7730 is based, the evaporative flow is decomposed into two components: transpiration secondary term that corresponds to evaporation inside the pores of the dry skin and evaporation from humid skin. Transpiration can be expressed without difficulty, but the physical details of the skin/confined air/clothing exchanges being complex, Fanger chose to statistically link the transpiration flow to metabolism and to skin temperature for an individual in a "comfort" situation. We should specify that statistical relations are the result of hundreds of tests in the climatic chamber.

The part of transpiration as water diffusion flux through the skin pores is expressed as:

$$q_{dif} = a_{dif}(p_{vsat}(T_{sk}) - p_{va}) \quad [3.12]$$

where P_v is water vapor pressure, p_{va} in ambient air, thus a function of temperature T_a and relative humidity φ, while $p_{vsat}(T_{sk})$ is the saturated vapor pressure at skin temperature. The coefficient a_{dif} is a "statistical" one: $a_{dif} = 3.05.10^{-3}$ W/m²Pa. The use of partial pressure rather than density is an old practice for the audience of experimenters who do not handle the transfer equations.

Function $p_{vsat}(T)$ can be derived with satisfactory precision within the range of temperatures of human environment from (Appendix 2):

$$log_{10}(p_{vsat}(T)) = 2.7858 + \frac{7.50T}{237.5+T} \quad [3.13]$$

where P_{vsat} is expressed in Pa and T in °C; $\varphi = p_v/p_{vsat}$.

The heat sweat flux is set as a linear function of skin temperature and is written:

$$q_{sw,r} = a_{sw}(Met - W - b_{sw});\ T_{sk} = a_{sk} - b_{sk}(Met - W) \quad [3.14]$$

For a physicist these relations are not obvious. We have to understand that when people in "thermal comfort" situations experience an increase in metabolism (activity) this has to be balanced by a decrease in air temperature, thus in skin temperature. The influence of clothing is hidden in the dispersion of point cloud, which statistically leads to setting the values of coefficients a_{sw}, a_{sk}, b_{sk}. This is sometimes denoted sweating flux "required" to ensure thermal comfort. We have:

$a_{sw} = 0.42$; $b_{sw} = 58.15$ W/m², $a_{sk} = 35.7$°C; $b_{sk} = 2.75.10^{-2}$ Km²/W (it is recalled that 1 K = 1°C).

Metabolism Met is thus reduced by the effort W made by an individual in his environment. Within the framework of a more physical model various formulations of evaporative exchanges can be proposed. According to the procedure for the calculation of sensible heat flux, it logically follows that we can write (q_{dif} included in q_{sw}):

$$q_{sw} = h_{evap} f_{cl} F_{pcl} \omega [p_{vsat}(T_{sk}) - p_{va})] \qquad [3.15]$$

F_{pcl} is the reduction factor due to clothing, which expressed by analogy is:

$$F_{pcl} = \left[1 + 0.155 I_{cl} f_{cl} i_v h_{evap}\right]^{-1} \qquad [3.16]$$

It should be understood that, on the one hand, we add up two vapor diffusion resistances, for clothing and boundary layer, and on the other hand we build in the clothing a diffusive resistance proportional to the thermal resistance. The dimensionless index i_v is a permeability index whose values are to be found in standard ISO 9920 or in specialized data tables.

h_{evap} is the vapor transfer pseudo-coefficient that by analogy with heat/mass transfers ($h_{cv} \rightarrow h_m$) is derived from the convective transfer coefficient following the Lewis relation:

$$h_{evap} = h_m \Delta h_v = a_{LW} h_{cv} \,;\, a_{LW} = 16.7.10^{-3} K/Pa \qquad [3.17]$$

Beware of the fact that mass transfer coefficient h_m is defined here for a vapor pressure difference and it will not be the same when a transfer coefficient defined with density gradient will be considered. Δh_v is the vaporization latent heat (differential enthalpy for phase transition). Attention should be given to the notation h_v for vapor enthalpy and h_{cv} for convective transfer coefficient.

Skin wettedness ω is the fraction of wet skin. Measurements indicate that ω = 0.06 corresponds to the above mentioned pore-evaporation. Starting from ω = 0.30 there is considerable discomfort and when ω = 0.50 sweat starts dripping down. The value of this parameter which sometimes serves as comfort criteria results from thermal equilibrium calculation. The sweating term plays a crucial role in certain circumstances and it will be derived from the thermal balance equation at equilibrium. We will come back to this term when more elaborate modeling will be discussed.

3.1.1.6. *Flux lost through respiration*

The thermodynamic expression would be:

$$q_{resp} = \dot{m}_{resp}(h_{v,out} - h_{v,in}) \qquad [3.18]$$

where $\dot{m}_{resp}$ is the respiratory mass flow rate and h_v is humid air enthalpy. Given the low importance of this term, Fanger simplifies the calculation and writes a sensible heat term and a latent heat term, considering in addition a linear dependence of the respiration flow rate with (metabolic) activity. We get:

$$q_{resp} = a_{1resp}Met(34 - T_a) + a_{2resp}(p_{vsat}(34°C) - p_{va}) \qquad [3.19]$$

where $a_{1resp} = 1.4.10^{-3}$/K; $a_{2resp} = 1.72.10^{-5}$/Pa; $p_{vsat}(34°C) = 5317$ Pa; T_a(°C).

Expired air is therefore considered saturated at the temperature of the nostrils' edge, that is at 34°C.

3.1.2. *Equilibrium balance equation in the Fanger model*

The (metabolic) gain equals various losses:

$$Met - W = q_{ray} + q_{cv} + q_{dif} + q_{sw,r} + q_{resp} \qquad [3.20]$$

expression to be considered in association with the equation for conduction through clothing in order to eliminate T_{cl} [3.6], [3.9].

With one exception, input data are the following parameters: Met, W, T_a, φ, T_p, V_a, I_{cl}, f_{eff}, ε. If, for example, we set $\varepsilon = 0.97$, $f_{eff} = 0.7$, $T_p = T_a$ and we define an ambient environment by T_a, φ, V_a we can choose a parameter of the individual Met or I_{cl} and deduce the other parameter in order to validate the thermal equilibrium. This simple equation can thus give rise to various representations of "thermal comfort" lines when one parameter varies: T_a, V_a, Met... It should be recalled that thermal comfort defined by physiological estimations has emerged only in the sweating expression "required" for comfort maintenance.

Let us move away from the previous state of thermal equilibrium by setting all the input parameters, without *a priori* satisfying the equilibrium equation. In proper heat engineering, the balance is:

$$\frac{Mc_p}{A_{sk}}\frac{dT}{dt} = Met - W - q_{ray} - q_{cv} - q_{sw} - q_{resp} \quad [3.21]$$

where M is the mass of the dressed body, A_{sk} is the skin surface area, c_p is the body's specific heat. The terms of the equation are algebraic, where q_{ray} can be either negative or positive.

Fanger uses the value of this equation to conceive an indicator of thermal comfort, called PMV, which is predicted mean vote, based on numerous tests in the climatic chamber:

$$PMV = (a_{1v}\exp(-a_{2v}Met) + a_{3v}) \\ \left(Met - W - q_{ray} - q_{cv} - q_{dif} - q_{sw,r} - q_{resp}\right) \quad [3.22]$$

where $a_{1v} = 0.303$ m²/W; $a_{2v} = 0.036$ m²/W; $a_{3v} = 0.028$ m²/W.

The PMV scale quantifies sensations: cold (-2), slightly cool (-1), warm (2), slightly warm (1), while 0 obviously indicates neutral. An acceptable environment for a work station will be characterized by $-0.5<PMV<0.5$. Moreover, PMV value allows us to deduce a (statistical) percentage of dissatisfied persons, denoted PPD (percentage of people dissatisfied) following the correspondence:

$$PPD = 100 - 95.exp[-(0.03353.PMV^4 + 0.2179.PMV^2)] \quad [3.23]$$

Let us note that neutrality defined by a PMV of zero accounts for 5% of dissatisfied people (Figure 3.2).

This model was completed in order to take into account the "air currents", that is to say turbulences induced by ventilation in a space. There are on the market stations for the measurement of comfort physical parameters: air temperature, wall radiation temperature, air velocity and turbulence, humidity. A small calculation program then gives the PMV for a typical subject, whose activity and clothing have been well defined.

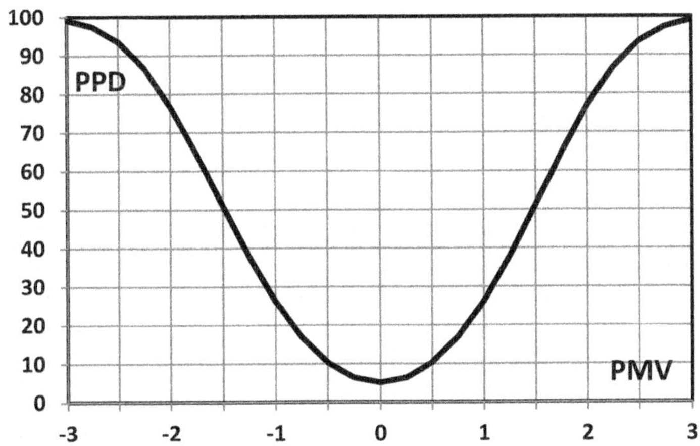

Figure 3.2. *Percentage of dissatisfied persons as a function of PMV index*

In order to globalize convective and radiative transfers that can be measured by a black globe, we often use operative temperature T_{op} [3.11], which allows accounting for the wall radiative influence.

To account for the wind effect, we will introduce an equivalent temperature, commonly used in weather forecast:

$$f_{cl}F_{cl}\left(h_{cv}+h_{ray}\right)\left(T_{sk}-T_{op}\right)=f_{cl}\mathring{F}_{cl}(\mathring{h}_{cv}+\mathring{h}_{ray})(T_{sk}-T_{eq}) \quad [3.24]$$

where the ° notation means in the absence of wind, with a radiative temperature equal to that of air (T_a).

The air currents cause a discomfort sensation linked to the air velocity V_a and fluctuations characterized by turbulence intensity T_u. This intensity is calculated as the standard deviation of fluctuations divided by mean velocity.

To account for this discomfort, a specific PPD is introduced and denoted DR (draught rate):

$$DR = (34 - T_a)(V_a - 0{,}05)^{0.62}(37V_aT_u + 3.14) \qquad [3.25]$$

where: V_a>0.05 m/s; T_a in °C; DR in %.

3.1.3. *Examples of ambient environment qualifications*

Parameters that characterize an ambient environment can easily be accessed and recorded with a measurement station comprising an anemometer (V_a, T_u), a thermo-hygrometer (T_a, φ) and a black globe ($\cong T_{op}$). The PMV criterion is conceived to qualify ambient environments in work premises or residential spaces. Using a pocket calculator or a spreadsheet, it is easy to program the fluxes and PMV calculation as a function of model input data M, I_{cl}, T_p, T_a, φ, $h_{cv}(V_a)$. Calculation can be performed in this order: T_{sk}, T_{cl}, p_{va}' p_{vsk}, q_{resp}, $q_{sw,r}$, q_{cv}, q_{ray}, PMV. To specify h_{ray}, iteration may be useful. Let us consider several examples under various conditions.

3.1.3.1. *Office work*

The units are the same as previously: h_{cv} in W/m²K, q and Met in W/m². The table below shows how air velocity and wall radiation (either warm or cold) influences comfort. V_a = 0.1 m/s → h_{cv} = 3.8 W/m²K; V_a = 0.4 m/s → h_{cv} = 7.6 W/m²K. For these chosen cases: Met = 90 W/m², I_{cl} = 1, φ = 0.5.

T_p(°C)	T_a(°C)	h_{cv}	q_{cv}	q_{ray}	$q_{sw,r}$	q_{resp}	PMV
20	20	3.8	21.5	32.1	25.3	8.1	0.1
20	20	7.6	34.5	25.5	25.3	8.1	-0.1
22	18	3.8	31.3	20.8	25.8	8.6	0.1
15	20	3.8	13.7	51.3	25.3	8.1	-0.3

Table 3.3. *Air velocity and wall radiation effect for an office*

An increase in wall average temperature allows for an equal decrease in air temperature. On the contrary, colder walls will cause discomfort.

We have denoted $q_{sw,r}$ the sweating flux "required" for neutral thermal comfort. When thermal balance with this flux is not zero (PMV is not zero) the real sweating flux is different.

3.1.3.2. *Comfort in the habitat of a person at rest*

Everyone has experienced transition from activity to rest, accompanied by a sensation of cold. The table below shows that because of a slowed down metabolism at rest, comfort requires better clothing or additional heat radiation. Met = 60 W/m², T_a = 20°C, φ = 0.5, h_{cv} = 3.5 W/m²K.

I_{cl}	T_p(°C)	q_{cv}	q_{ray}	$q_{sw,r}$	q_{resp}	PMV
1	20	21.4	34.8	13.5	5.4	-0.9
1.5	20	16.6	26.8	13.5	5.4	-0.1
1.5	22	20.3	19.	13.5	5.4	0.1

Table 3.4. *Comfort of a person at rest*

3.1.3.3. *Performance hall*

Here are two cases of ambient environment of a performance hall, one with low occupancy rate, the other at maximum occupancy, which generates strong radiation in the proximity. Met = 60 W/m², I_{cl} = 1, φ = 0.5, h_{cv} = 3.8 W/m²K.

T_p(°C)	T_a(°C)	q_{cv}	q_{ray}	$q_{sw,r}$	q_{rep}	PMV
23	23	17.8	27.2	12.7	4.9	-0.2
27	19	37.6	4.1	13.7	5.6	-0.05

Table 3.5. *Comfort in a performance hall*

3.1.3.4. *Swimming pool*

Let us consider a sitting, dry bather at rest in the swimming pool premises. The bather is subjected to high air temperatures but should a cold radiation emerge (big glazed areas) discomfort is guaranteed. Met = 60 W/m², I_{cl} = 0, T_a = 28°C, φ = 0.8, h_{cv} = 3.8 W/m²K.

T_p(°C)	q_{cv}	q_{ray}	$q_{sw,r}$	q_{rep}	PMV
28	22.7	36.9	7.8	2.8	-0.6
20	22.4	82.6	7.8	2.8	-3.3

Table 3.6. *Comfort in a swimming pool*

3.1.3.5. *Several particular outdoor situations*

Let us consider a skier in two configurations: skiing up (Met = 240 W/m²) and down (Met = 120 W/m²). Thermal adaptation is to be performed by the clothing, as for the same wear the conclusion is: too warm when skiing up and too cold when skiing down. $I_{cl} = 2$, $T_p = T_a = 0°C$, $\varphi = 0.5$.

Met	h_{cv}	q_{cv}	q_{ray}	$q_{sw,r}$	q_{rep}	PMV
240	17	66.1	17.7	87.7	31.9	1.2
120	40	88.8	10.0	39.9	15.9	-1.2

Table 3.7. *Comfort for a skier*

Let us consider a (dry) bather on the beach in summer. Depending on the "air currents" or on the ambient radiation (direct or sand–diffused sunlight) the bather can experience the transition from cold to warm sensation. Met = 60 W/m², $I_{cl} = 0$, $T_a = 30°C$, $\varphi = 0.7$.

T_p(°C)	h_{cv}	q_{cv}	q_{ray}	$q_{sw,r}$	q_{rep}	PMV
30	5	19.9	24.9	8	2.7	0.3
30	10	39.5	24.7	8	2.7	-0.9
40	10	40.5	-39.1	8	2.7	2.9

Table 3.8. *Comfort situations on the beach*

3.1.3.6. *Marathon runner*

An article published in a medical journal [CAB 97] suggests that the soldier/runner who ran from Marathon to Athens with the good news of victory over the Persians (in 490 BC) and died shortly after delivering the message, may have died due to the brain overheating after sudden interruption of the wind, when the race stopped. Let us try to quantify the possible thermal exchanges. Being outside of the validation range, the Fanger model is not applicable, but a certain number of elements can be retained. The calculation of convection, radiation and respiration fluxes is still valid. Sweating required for thermal equilibrium makes no sense here and we will replace it by a resulting sweat in order to have a steady balance, which is the case for such a long race:

$$q_{sw} = Met - q_{ray} - q_{cv} - q_{resp}$$

A skin temperature should be *a priori* fixed and this should be very near $T_{sk} = 35°C$. Based on the value of q_{sw} a value for ω will be derived through

[3.15]–[3.17], which will serve as evaluation criterion. An analysis of results can be performed on the table below (i_v = 0.14). Met = 400 W/m²; I_{cl} = 0.1; φ = 0.5; T_{sk} = 35°C:

T_p	T_a	h_{cv}	q_{ray}	q_{cv}	q_{resp}	q_{sw}	ω
25	25	24	29.8	169.3	30.4	170.5	0.11
30	30	24	15.3	84.5	24	276.2	0.20
30	30	5	19.4	22.2	24	334.4	1.13

Table 3.9. *Thermal equilibrium of the Marathon runner*

The first two lines indicate a possible balance during the race (h_{cv} = 24 W/m²K) with acceptable fractions of wetted skin. Values $0.05<\omega<0.5$ are admitted. After the sudden stop of the race, the body inertia is such that metabolism continues to operate for at least a dozen minutes and the equilibrium balance at low ventilation (h_{cv} = 5 W/m²K) is not compatible with reality anymore, and this translates as overheating of the body, particularly of the brain, and consequent death.

3.1.3.7. *Equivalent temperature without wind*

The Fanger model allows for the deduction of an equivalent temperature T_{eq} also named "felt temperature" equation [3.24], factor F_{cl} given [3.10]. This temperature obviously supposes a convention on the individual's clothing and on the convective exchange coefficient of the reference situation. Let us consider an individual who takes a good walk outdoors during winter (M = 240 W/m²) at two possible temperatures (0 and –10°C) with $\mathring{h}_{cv}$ = 3.5 W/m²K, φ = 0.5. Let us test the influence of air velocity (relative velocity) of 1, 2, 3 m/s, for which h_{cv} is 12.1; 24.2; 36.3 W/m²K.

I_{cl}	$T_p = T_a$	h_{cv}	PMV	T_{eq}
1.5	0	12.1	0.56	-7.0
1.5	0	24.2	0.30	-10.4
1.5	0	36.3	0.17	-11.9
2	-10	12.1	0.23	-17.7
2	-10	24.2	0.01	-21.1
2	-10	36.3	-0.10	-22.5

Table 3.10. *Felt temperature*

3.2. Gagge model

3.2.1. *A simple, unsteady and regulated geometrical model*

Making a step further in the simulation of human thermal behavior, Gagge [GAG 71] perfects the "potato" model, which is now composed of core and skin and a minimum blood exchange between the two, so that heat transfer through blood flow regulated by vasodilatation (increase of blood flow) or vasoconstriction (reduction of blood flow) can be introduced. Thermal exchanges introduced in the Fanger model are still valid: convection, radiation, respiration and transpiration, the latter being the second regulation parameter (Figure 3.3).

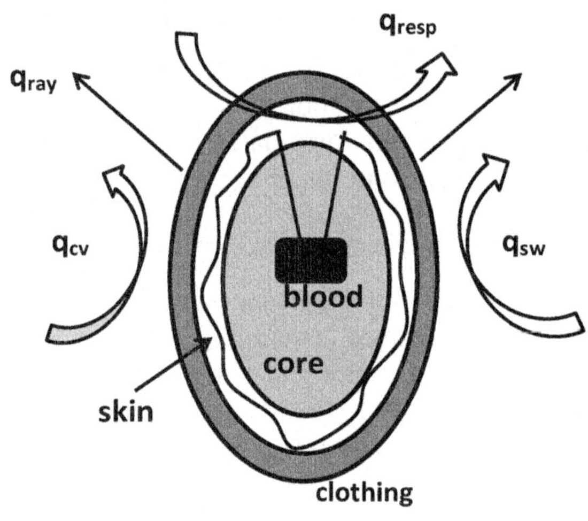

Figure 3.3. *Gagge model*

Similarly to the Fanger model, in this model clothing filters heat and vapor and has no effect on thermal mass. The "man" unsteady system has therefore two components: skin and core (model with two "nodes"). Let M, c_p, T be mass, specific heat and temperature respectively, with indices sk for skin and cr for core. Thermal balance equations are:

$$\left(\frac{Mc_p}{A}\right)_{sk}\frac{dT_{sk}}{dt} = -q_{cv} - q_{ray} - q_{sw} + q_{bl} + q_{conduc} \quad [3.26]$$

$$\frac{M_{cr}c_{p,cr}}{A_{sk}}\frac{dT_{cr}}{dt} = Met - W - q_{resp} - q_{bl} - q_{conduc} \qquad [3.27]$$

There are two new terms, q_{bl} which is the heat flux carried by blood from core to skin and $q_{conduct}$ which is the conduction flux from core to skin. Although fluxes are algebraic, the signs introduced favor current cases. Let us note that heat production through metabolism takes place entirely in the core (muscles). Similarly to the Fanger model, fluxes are divided by skin surface area A_{sk}. This model greatly simplifies the complexity of blood and conductive exchanges and we can write:

$$q_{bl} = \frac{\dot{V}_{bl}c_{p,bl}}{A_{sk}}(T_{cr} - T_{sk}) \qquad [3.28]$$

$$q_{conduct} = K(T_{cr} - T_{sk}) \qquad [3.29]$$

$\dot{V}_{bl}$ is the blood volumetric flow rate and K is the coefficient that globalizes core/skin conduction. The delicate point of transpiration is treated here in a more physical manner. Sweating flow rate $\dot{m}_{sw}$ is imposed through internal regulation by temperature difference signals, as we will see further on, and this flow rate determines a ratio ω of wetted surface as introduced above [3.15]–[3.17]. Vapor diffusion on dry skin (through the pores) is taken into account here by imposing a minimum of 0.06 for ω. Beyond that we decompose as previously $\omega = 0.06+\omega_r$ and $q_{sw} = q_{dif}+q_{sw,r}$ except that this is not a comfort model and that the transpiration flow rate calculated is not "required" for comfort, but imposed through a physiologist statistical relation.

The relation set between evaporated flux and sweating flow rate introduces a command by two temperature differences $\varepsilon_{sk} = T_{sk}$-34.1 and $\varepsilon_{cr} = T_{cr}$-36.6 as follows:

$$q_{sw,r} = a_{sw1}\dot{m}_{sw,r}(2)^{\varepsilon_{sk}/3} \qquad [3.30]$$

$$\dot{m}_{sw,r} = a_{sw2}\varepsilon_{cr} + a_{sw3}\varepsilon_{sk}\varepsilon_{cr} \qquad [3.31]$$

where $\varepsilon_{cr} \geq 0$, $\varepsilon_{sk} \geq 0$. Reference temperatures of 34.1 and 36.6 are average temperatures for comfortable thermal equilibrium at rest. The cold or warm "signals" issued by the skin or core impact also vasoconstriction or vasodilatation by the intermediary of blood flow rate:

$$\begin{cases} \dot{V}_{bl} = (a_{bl1} + a_{bl2}\varepsilon_{cr});\ \varepsilon_{cr} \geq 0 \\ \dot{V}_{bl} = \frac{a_{bl1}}{1-a_{bl3}\varepsilon_{sk}};\ \varepsilon_{sk} \leq 0 \end{cases} \qquad [3.32]$$

The system to be solved is thus composed of two unsteady equations [3.26], [3.27] with two unknown T_{sk}, T_{cr} with the precision equations of various terms [3.15]–[3.17], [3.28]–[3.32] as well as the initial definition equations [3.1], [3.5], [3.9], [3.10], [3.11] and [3.19] but with a skin temperature that is not a function of metabolism. The programming for the numerical resolution is quite easy on various pocket calculators or on a spreadsheet. What we want is therefore to test the adaptation of the human individual to particular situations where a parameter suddenly varies (effort, temperature, humidity, etc.). The criteria for estimating "cold" and "warm" are quite simplified with the differences of temperature ε, and furthermore in a warm environment ω_r is a significant criterion. Finally, this quite simple model nevertheless requires valid physiological coefficients, which are difficult to obtain.

3.2.2. *Response of "human system" to a sudden change in metabolism*

Let us consider a quite common experience: starting from a given equilibrium state (clothing, environment and metabolism) a change of activity occurs, and this requires an increase in metabolism. How does the increase in temperature take place? Let us state the data.

Let us consider an individual whose weight is 81.7 kg (M_{sk} = 3.4 kg; M_{cr} = 78.3 kg) and skin surface area A_{sk} = 2 m², with thermal properties given by $c_{psk} = c_{pcr}$ = 3,492 J/kgK; c_{pbl} = 4.187.10^6 J/m^{3}K; K = 5.28 W/m²K.

The basic blood flow rate is defined by a_{bl1} = 3.79.10^{-6} m^3/s and by the two regulation parameters a_{bl2} = 41.66.10^{-6} m$^{3/}$sK; a_{bl3} = 0.5/K.

The transpiration flow rate is defined by the constants a_{sw1} = 2.52*10^6 J/kg; a_{sw2} = 6.94*10^{-5} kg/m²Ks; a_{sw3} = 2.78.10^{-5} kg/m²K²s.

The quantities f_{cl}, F_{cl}, and F_{pcl} are defined as previously with i_v = 0.143.

The parameters of the individual's initial state are: Met = 120 W/m², I_{cl} = 1, $T_a = T_p$ = 20°C, $h_{cv} = h_{ray}$ = 5 W/m²K, φ = 0.5.

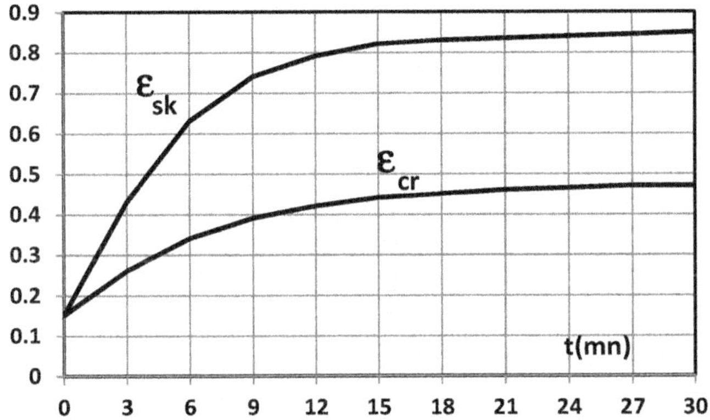

Figure 3.4. *Gagge model. Response to a jump in metabolism from 120 to 240 W/m²*

The numerical method of resolution of the system of two equations [3.26], [3.27] will determine the initial state of equilibrium by the convergence of the launched calculation with a couple of arbitrary values (T_{sk}, T_{cr}). A metabolism which has doubled its value (240 W/m²) is imposed at the initial time, and this determines an increase in temperatures and sweating in the defined environment (Figures 3.4 and 3.5).

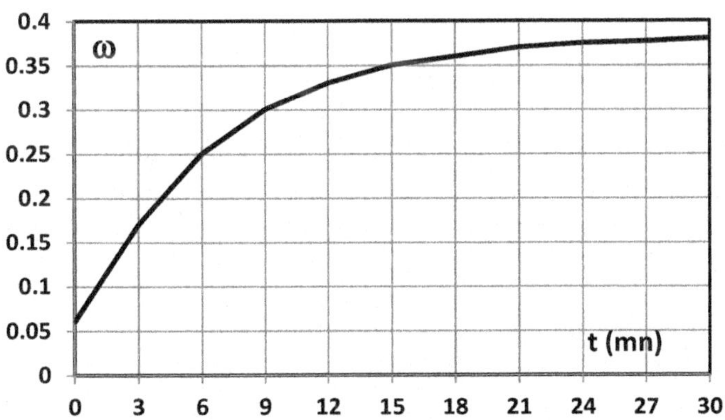

Figure 3.5. *Sweating response to a jump in metabolism*

The system reaches a steady state in a characteristic time which is about half an hour. This quite simple model is credited with highlighting the time constants of the individual and thus with predicting the "bearable" discomfort, for example starting from parameter ω, ratio of wet skin, for difficult professional situations (firemen, iron and steel industry, etc.).

3.3. Stolwijk 25 node model [STO 70, STO 71]

This is a much more elaborate model, as it "divides" the human body into six segments (head, trunk, legs and feet) and 4 layers per segment (core, muscles, fat and skin), which gives 24 "nodes", and a blood pool which supplies skin and muscles (Figure 3.6). Complexifying a model offers the advantage of fine evaluation of responses, but the disadvantage is that a large number of parameters, in this case physiological, each of them affected by uncertainty, need to be entered. As previously, the system is regulated by blood flow (vasodilatation and vasoconstriction), transpiration and shivering, corresponding to additional muscle/metabolic activity in case of cold sensation. For each layer i of a segment or "node", assigned with average quantities, an energy balance equation can be written:

$$M_i c_{p,i} \frac{dT_i}{dt} = Met_i + q_{conduct,i} - q_{cv,i} - q_{ray,i} - q_{sw,i} - q_{resp} + q_{bl,i} \quad [3.33]$$

$$q_{conduct,i} = \sum_i K_{i,j}(T_j - T_i)A_i \quad [3.34]$$

$$q_{bl,i} = \alpha_i \dot{m}_{bl,i} c_{p,bl}(T_{CB} - T_i) \quad [3.35]$$

In this case, thermal flux terms are not considered on m^2 of skin. Surface areas will thus be factored in most of the terms. $K_{i,j}$ is the conductance coefficient between layer i and an adjacent layer j. The "central blood pool" supplies blood flows as follows:

$$M_b c_{p,bl} \frac{dT_{CB}}{dt} = -\sum_{i=1}^{24} q_{bl,i} \quad [3.36]$$

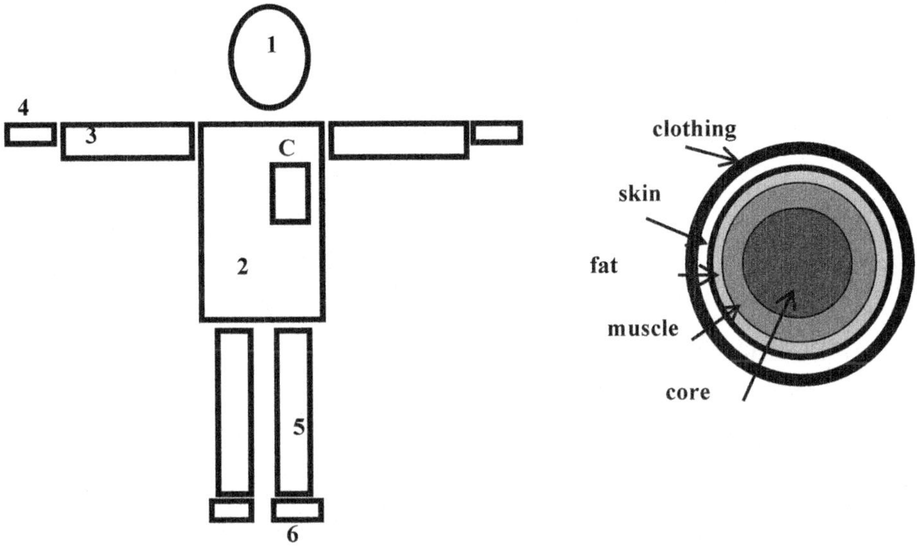

Figure 3.6. *Division according to Stolwijk model*

Heat fluxes are algebraic quantities, even though the signs set offer an *a priori* interpretation.

While solving 25 ("more or less" linear) equations with 25 unknowns does not pose great difficulties for scientists today, "feeding" the model with physiological and physical coefficients is an arduous task. What convective and radiative transfer coefficients should be considered on a leg? What blood distribution should be considered for a hand? and the list of questions is very long.

The human system's control loop is now more complex. Each compartment is fitted with a thermal receiver, which sends a "warm" or "cold" signal in comparison with a reference temperature. The set of signals is weighted by control techniques for an action instruction, either transpiration and vasodilatation, or shivering and vasoconstriction.

Based on this principle of the model, a version can be adapted for a precise purpose, the "division" can be modified into more or fewer elements, and a more detailed clothing model can be associated (Chapter 4). We will see an example of a thermal model of a baby.

3.4. Thermal model of a baby lying down

This study is motivated by concerns over sudden infant death. For a long time the medical profession has pointed the finger at the over-insulation of babies during sleep as the main cause of this pathology, and advises young parents to act with "moderation" in view of overnight thermal protection: clothing, blanket, room temperature, and also recommends putting the baby to sleep on the back or on one side, rather than on the stomach.

3.4.1. *Geometrical division*

This context allows for a simplified geometry of the thermal model "Stolwijk style": a spherical head composed of skin and one core, and a set of trunk-legs-arms as a cylinder composed of skin, a layer of muscles-fat and a core (Figure 3.7).

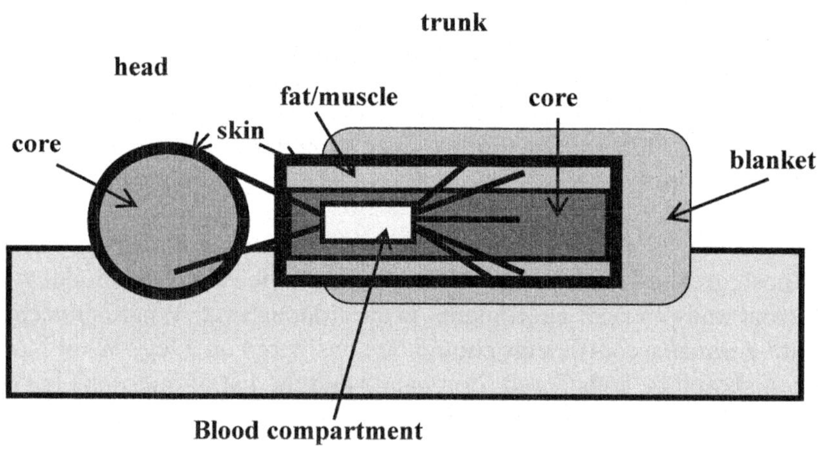

Figure 3.7. *Division of the baby model*

Statistical weight and size data of a 3-month baby (age at which risk of sudden death peaks) lead at setting the dimensions as follows: diameter of the spherical head and cylindrical trunk is 13.4 cm; length of trunk is 37 cm; skin thickness is 2 mm; fat-muscle layer thickness is 2.5 cm. Various weights are deduced based on an average density of 0.94 g/cm^3 just as the heat capacity (Mc_p) of each compartment as indicated in the table below:

	Head skin (h, sk)	Head core (h, co)	Trunk skin (tr, sk)	Trunk muscles (tr, m)	Trunk core (tr, co)	Blood Compartment (bl, cp)
Volume (cm^3)	106.2	1,150	302.2	3,051	1,859.8	
Mass (g)	99.8	1,081	264.1	2,868	1,268.2	480
C_p (J/kgK)	2,800	3,000	2,800	2,400	3,000	4,187
Mc_p (J/K)	279.4	3,243	739.5	6,883	3,804.6	2,009.7

Blood weight was deducted from that of the core-trunk. Total weight is M = 6.061 kg.

The system of equations [3.33]–[3.35] now refers to six compartments (6 "nodes") but given that the baby is lying down, thermal interaction with the support-mattress imposes at least one additional node. Let us note on this example that the very "art" of modeling consists of homogeneous simplification of various aspects of the problem posed. More advanced modeling might consider 3D aspects of heat diffusion in the mattress, but due to preliminary "engineer-like" evaluation of heat exchanges between the baby and his environment we know that here conduction in the mattress is a very secondary flux. Let us now specify the various fluxes.

3.4.2. *Metabolism and respiration*

Metabolism is a function of baby weight and of an "overheating" coefficient that depends on the central temperature of the trunk [JAR 92]:

$$Met = Met_b M f_{met};\ Met_b = 2.4\ W/kg;\ f_{met} = 2.5^{(T_{tr,co}-37)/10};\ T_{tr,co} \geq 37°C \quad [3.37]$$

This global metabolism is distributed as follows:

$$Met_{h,co} = 0.4\ Met;\ Met_{tr,m} = 0.10 Met;\ Met_{tr,co} = 0.5\ Met \quad [3.38]$$

Respiration is expressed by a difference of enthalpy of the saturated humid air that goes in and out, at the head-skin temperature:

$$q_{resp} = \rho_a \dot{V}_{resp}\left[h_{sat}\left(T_{h,sk}\right) - h_a(T_a, \varphi_a)\right] \quad [3.39]$$

where ρ_a is the density of ambient air, $\dot{V}_{resp}$ is the average respiratory flow rate, h is air enthalpy that can be derived from the following system (Appendix 2):

$$\begin{cases} h = c_{pa}T + W_\varphi(\Delta h_v^o + c_{pv}T);\ c_{pa} = 1006J/kgK;\ c_{pv} = 1860J/kgK \\ \Delta h_v^o = 2.5016.10^6 J/kg \\ W_\varphi = a\frac{\varphi p_{vsat}(T)}{p-\varphi p_{vsat}(T)}; p = 1.013.10^5 Pa; a = 0.622 \\ log_{10}(p_{vsat}(T)) = 2.7858 + \frac{7.5T}{237.3+T}; T(°C) \end{cases} \quad [3.40]$$

For a 3-month baby (6 kg) the representative figures are: respiratory volume 9 ml/kg, 40 respirations per minute or an average respiratory flow rate $\dot{V}_{resp}$ = 3.6.10^{-5} m^3/s. The respiration rate is distributed between trunk and head cores as follows:

$$q_{res,tr,co} = 0.6\ q_{resp};\ q_{resp,h,co} = 0.4\ q_{resp}$$

3.4.3. *Exchanges of the uncovered part of the head*

Head is the main source of heat dissipation because of the direct contact between skin and air. A natural convection mechanism is set up. Thermal boundary layers touch upon the free surface of the head-sphere and give rise to a hot plume that emerges above and can be viewed by luminous planes. This thermal draft boundary layer is characterized by a gradient of temperature T and of vapor density ρ_v. If c_o is the coefficient of the head free surface, then thermal convection and evaporative fluxes can be written:

$$q_{cv,h} = h_{cv}(T_{h,sk} - T_a)\ c_o\ A_h \quad [3.41]$$

$$q_{evap,h} = k\ (\rho_{vsat}(T_{h,sk}) - \varphi_a\ \rho_{vsat}(T_a))\ \omega\ c_o\ A_h\ \Delta h_v \quad [3.42]$$

where A_h is the head surface area, ω humid surface ratio due to sweating, with ω at least 0.06 (diffusion through skin) and always lower than 1. h_{cv} and k are heat and mass transfer coefficients. They can be derived one from the other by analogy with boundary layer equations, named Chilton–Colburn analogy, which for laminar flow can be written:

$$Sh = Nu\left(\frac{Sc}{Pr}\right)^{1/3}; Sh = \frac{kL}{D_v}; Nu = \frac{h_{cv}L}{\lambda}; Sc = \frac{\upsilon}{D_v}; Pr = \frac{\mu c_p}{\lambda} \quad [3.43]$$

where Sh is Sherwood number, Nu is Nusselt number, Sc is Schmidt number, and Pr is Prandtl number. L is a length characteristic to the problem. The ratio k/h_{cv} is derived with the following values of physical parameters. For air and water vapor in the air at around 20°C, $\lambda = 0.025$ W/mK; $D_v = 2.6.10^{-5}$ m²/s; Pr = 0.72; Sc = 0.61 or $k/h_v = 0.99.10^{-3}$ K/Pa. To make the connection with Lewis analogy [3.17], let us compare definitions:

$$q_{sw} = a_{Lw}\, h_{cv}\, (p_{vsat}(T_p)\text{-}p_{va}) \text{ and } q_{sw} = k\, (\rho_{vsat}(T_p)\text{-}\rho_{va})\, \Delta h_v$$

which together with the perfect gas equation for water vapor around an average temperature in the boundary layer $\overline{T}$ gives:

$$a_{LW} h_{cv} = k \frac{\Delta h_v}{r_v \overline{T}} \qquad [3.44]$$

From data $a_{LW} = 16.7.10^{-3}$ K/Pa; $r_v = 461.4$ J/kgK; $\overline{T} = 300$K; $\Delta h_v = 2.5.10^6$ J/kg, we get: $k/h_v = 0.92.10^{-3}$ K/Pa.

These two analogies are thus based on the same similarity between the equations, and the result is likely to vary a little depending on physical context (turbulence in particular).

The radiation of the uncovered part of the head will be expressed in the usual manner:

$$q_{ray,h} = h_{ray}\, F\, (T_{h,sk}\text{-}T_a)\, c_o\, A_h \qquad [3.45]$$

where T_a is the air and wall temperature, F is a shape factor for reducing the radiative surface area that takes into account the environment surrounding the bedding (pillow, etc.) and is a matter of estimation based on solid angles.

3.4.4. *Conduction between body layers*

Given a good order of magnitude of thermal conductivity of the body (λ = 0.5 W/mK), the problem posed is to determine the conductive heat flux between two layers of the head or trunk. The nodal method principle is the division into fine layers of uniform temperature, so that inter-layer exchanges can be considered steady. The division performed here is quite rudimentary, but in line with the result sought concerning the skin and the muscle cylinder, but much less for the cores. This is why two methods for writing conduction will be used. The steady heat flux between layers of skin and muscles of the cylindrical trunk of length L can be written as:

$$q_{1,2} = \frac{2\pi\lambda L}{\ln\left(\frac{r_2}{r_1}\right)}(T_1 - T_2) \qquad [3.46]$$

Radius vectors r_1, r_2 of the "nodes" are not exactly at the center of the cylindrical layer, but divide each layer into 2 under-layers of the same volume (r_1 = 5.4 cm; r_2 = 6.6 cm). A global conduction coefficient ahead of the temperature difference, and comprising the surface area, will be denoted $K_{sk\text{-}tr,mu}$ = 5.79 W/K.

The conduction coefficient between the cores and the adjacent layer is obtained by considering the unsteady analytical solution of conduction in a sphere or a cylinder in comparison with the nodal solution. In effect, nodal modeling consists of setting the heat equation for the core under the form:

$$C\frac{dT}{dt} = K(T_p - T); T(t); C = \varrho c_p V \qquad [3.47]$$

whose solution expressed in terms of flux with $T(r, 0) = T_{in}$, $T(r_f, t) = T_p$ is:

$$\frac{q(t)}{q_\infty} = 1 - exp\left(-\frac{K}{C}t\right) \qquad [3.48]$$

K is to be determined by comparison with the analytical solution T(r, t) of:

$$\varrho c_p \frac{\partial T}{\partial t} = \lambda \frac{1}{r^i}\frac{\partial}{\partial r}\left(r^i \frac{\partial T}{\partial r}\right); cylinder\ i = 1; sphere\ i = 2 \qquad [3.49]$$

whose solution for a cylinder of radius r_f is:

$$\frac{q(t)}{q_\infty} = 1 - \sum_{n=1}^{\infty}\frac{4}{(r_f\alpha_n)^2} exp\left(-\frac{\lambda}{\rho c_p}\alpha_n{}^2 t\right) \qquad [3.50]$$

Coefficients $r_f\,\alpha_n$ are the roots of the Bessel functions $J_0(r_f\,\alpha_n) = 0$. The first three are sufficient for the required precision.

n	$r_f\,\alpha_n$
1	2.40482
2	5.52008
3	8.65373

Table 3.12. *Roots of the Bessel function*

The solution for the sphere is:

$$\frac{q(t)}{q_\infty} = 1 - \frac{6}{\pi^2}\sum_{n=1}^{\infty}\frac{1}{n^2}exp\left[-\frac{\lambda}{\rho c_p}\left(\frac{n\pi}{r_f}\right)^2 t\right] \quad [3.51]$$

Three terms will be sufficient for the precision. The values of K are “chosen” at best to obtain a good concordance in the targeted time frame as shown in Figure 3.8 for the sphere-core, $K_{sk\text{-}h,co}$ = 1.9 W/K and $K_{m\text{-}tr,co}$ = 4 W/K.

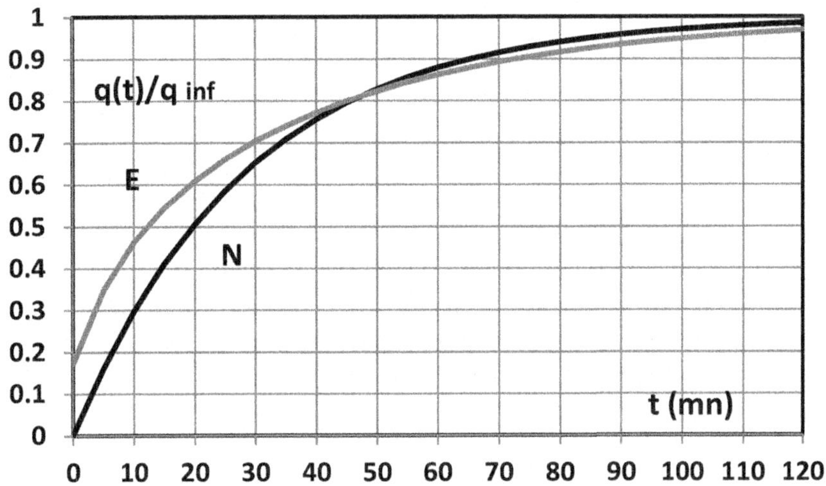

Figure 3.8. *Conduction in the sphere. Nodal equation (N) with K = 1.9 W/K and exact equation (E)*

3.4.5. *Sensible heat exchanges of the trunk*

Thermal resistance of clothing and blankets is obviously a key factor of overheating. A comprehensive study [BAC 91] of thermal resistances (R_{th}) depending on types of baby bedding is available and we will retain a wide range from 0.5 to 2 m²K/W. The conduction flux of the trunk has to be decomposed in an external part $q_{sens,tr}$, corresponding to the outer face of the bedding which exchanges through convection and radiation with the surrounding surfaces and the inner part which diffuses into the mattress q_{mat}.

The first flux is expressed analogously to equations [3.9] and [3.10] as it is still a matter of summing up thermal resistances:

$$q_{sens,tr} = (h_{ray}F + h_{cv})F_{cl}(T_{tr,sk} - T_a)c_o A_v \qquad [3.52]$$

$$F_{cl} = \left[1 + f_{cl}R_{th}(h_{ray}F + h_{cv})\right]^{-1};\ f_{cl} = A_v/A_{tr,sk} \qquad [3.53]$$

A_v is the outer surface area of the blanket (Figure 3.7), $A_{tr,sk}$ is the surface area of the trunk skin, F is a shape factor of radiation, which is *a priori* of the same order as for the head.

The detailed evaluation of the conduction flux in the mattress is very complex, but given its secondary importance it can be simplified as:

$$q_{mat} = K_{sk,tr-mat}(T_{tr,sk} - T_{mat})(1 - c_o)A_v \qquad [3.54]$$

The conduction coefficient introduced has to take into account the thermal resistance of clothing and of a part of the mattress (up to the node position). Thermal conductivity coefficients of the bedding fabric and of the mattress foam being known, thermal resistances are calculated based on thicknesses. For example, a common cotton fabric has $\lambda = 0.042$ W/mK and a mattress foam has $\lambda = 0.03$ W/mK.

3.4.6. *Trunk evaporation*

The detailed problem of trunk evaporation is quite complex and it will be simplified, given its reduced contribution to the heat balance. Thus, we will neglect the condensation, at the start of the process, of a part of vapor in the clothing, which is a heat source. Baby's clothing "filters" the vapor flux and formulation [3.15], [3.16] becomes:

$$q_{evap,tr} = k\, F_{pcl}\, \omega\left(\rho_{vsat}\left(T_{tr,sk}\right) - \rho_{v,a}\right)c_o\, A_v \Delta h_v \qquad [3.55]$$

The driving gradient is in this case a difference in vapor density, and not a vapor pressure; k is the "vapor" exchange coefficient on the outer face of bedding, ω the ratio of wet skin surface area and F_{pcl} is the "vapor filter" coefficient of the clothing + bedding thickness:

$$F_{pcl} = \left[1 + f_{cl}\, R_{th,tr} \mathrm{L_w}\, i_v\, k\right]^{-1} \qquad [3.56]$$

The ratio k/h_{cv} is derived from [3.43]. Coefficient i_v is obtained from databases with measured values of thermal conductivity coefficients and vapor diffusivity [HAV 90, HAV 99] in textile layers. The definition of i_v leads to that of L_W. In effect, let us consider a textile layer of thickness e_p for which a thermal resistance R_{th} and an evaporative resistance R_{evap} are measured:

$$\begin{cases} R_{th} = \frac{e_p}{\lambda};\ R_{evap} = \frac{e_p}{D_v} \Rightarrow R_{evap} = R_{th}\frac{\lambda}{\lambda^o}\frac{D_v^o}{D_v}\frac{\lambda^o}{D_v^o};\ L_W = \frac{\lambda^o}{D_v^o} \\ R_{evap} = R_{th}L_W i_v;\ i_v = \frac{\lambda}{\lambda^o} / \frac{D_v}{D_v^o} \end{cases} \quad [3.57]$$

The constant L_W, analogous to the Lewis constant previously introduced, is related to mass transfer coefficients expressed with vapor density gradients and consists of a ratio of reference data chosen for a motionless layer of humid air. For example, for current temperatures and pressures $\lambda^o = 2.6.10^{-2}$ W/mK; $D^o_v = 2.6.10^{-5}$ m²/s, we have $L_W = 10^3$ J/m³K. For orders of magnitude of i_v, current values indicated in the literature are: $\lambda = 0.04$ W/mK ($\lambda > \lambda^o$); $D_v = 1.3.10^{-5}$ m²/s ($D_v < D^o_v$), which gives $i_v = 3.08$. Let us note that thermal conductibility can increase by 50% in case of bound water saturated textile (adsorbed water vapor).

3.4.7. *Blood convection*

In a hot environment the network of superficial veins has a preferential activity (vasodilatation) and blood flow dissipates internal heat. This phenomenon is similar to a counter flow heat exchanger whose model we can consider for each compartment i with the expression:

$$\dot{q}_{blood,i} = \dot{m}_{blood,i}c_{p,blood}(T_C - T_i) \quad [3.58]$$

The notion "blood compartment" pertains rather to a "model" than to physiology and it does not exactly cover the heart. Blood flow rates through each compartment are certainly not measurable, but they are deduced from a certain amount of data. They are expressed from a basic flow rate $\dot{m}_{blood,b} = 4.5.10^{-5}kg/s$ [SUY 73] and the maximum value for vasodilatation $\dot{m}_{blood,max} = 20\dot{m}_{blood,b}$. The flow rate variation with core temperature is expressed as a linear variation based on a temperature difference:

$$\dot{m}_{blood} = \dot{m}_{blood,b} + \frac{T_n - 36.8}{\Delta T_n}\Delta\dot{m}_{blood};\ \Delta T_n = 0.7°C;\ \Delta\dot{m}_{blood} = 19\dot{m}_{blood,b} \quad [3.59]$$

where T_n is a "core" temperature (°C), defined by $T_n = \max(T_{h,co}, T_{tr,co})$.

The blood flow rate difference results from an estimation of maximum flow rate for an adult as being 20 times the basic flow rate. The flow rate "regulator" instruction is thus applicable for temperatures between 36.8 and 37.5°C. When T_n<36.8°C $\dot{m}_{blood} = \dot{m}_{blood,b}$, and when T_n>37.5°C $\dot{m}_{blood} = \dot{m}_{blood,max}$.

The various model compartments receive a fraction of this blood flow rate, the skin flow rates being regulated, while the muscle and core flow rates are constant:

$$\begin{cases} \dot{m}_{blood,h,sk} = 0.3\dot{m}_{blood};\ \dot{m}_{blood,tr,sk} = 0.7\dot{m}_{blood} \\ \dot{m}_{blood,h,co} = 3\dot{m}_{blood,b};\ \dot{m}_{blood,tr,co} = 7\dot{m}_{blood,b};\ \dot{m}_{blood,m} = 1.5\dot{m}_{blood,b} \end{cases} \quad [3.60]$$

3.4.8. *System of equations*

The equations of 6 compartments can be written as:

$$M_i c_{p,i} \frac{dT_i}{dt} = Met_i + q_{conduc,i} - q_{cv,i} - q_{ray,i} - q_{evap,i} - q_{resp} + q_{blood,i} \quad [3.61]$$

This is a general expression, and it is obvious that certain fluxes are zero through certain compartments. We should add to these 6 equations the one for "mattress-node":

$$\left(Mc_p \frac{dT}{dt}\right)_{mat} = K_{sk,h-mat}(T_{h,sk} - T_{mat})(1 - c_o)A_h + K_{tr,sk-mat}(T_{tr,sk} - T_{mat})(1 - c_o)A_v - K_{mat,a}(T_{mat} - T_a)(1 - c_o)(A_h + A_v) \quad [3.62]$$

Conduction coefficients are estimated depending on the type of material (mattress) and thicknesses. For example, for a node situated in the middle of a foam mattress (λ = 0.03 W/mK) of thickness 10 cm, the coefficients K are of about 0.6 W/m²K. Taking into account the measurements of the "hot spot" induced in the mattress by the head and the body, we set a thermal mass for a surface area of 50cm*39 cm, with a depth of 3 diameters underneath the body, that is 1.95 kg, $(Mc_p)_{mat}$ = 2,340 J/kg (ρ = 100 kg/m^3; c_p = 1,200 J/kgK).

System [3.61], [3.62] admits a classical resolution: discretization and implicit method equation:

$$C_i \frac{T_i^{t+\delta t} - T_i^t}{\delta t} = \sum_j a_{i,j} T_j^{t+\delta t} \qquad [3.63]$$

The system of linear equations is solved by matrix inversion. A number of programs can be used for this purpose. Using an advanced spreadsheet is also a possibility. Coefficients $a_{i,j}$ may be a function of temperature, which will require an iteration at each time step.

3.4.9. *Simulation results*

What results are to be expected from such a model? How does a baby's temperature increase and what are the key parameters of this process?

A prerequisite for answering these questions is testing the coherence of the parameters introduced. In effect, given the model's large number of physical and physiological parameters, we should check that they are in "agreement" for several situations of plausible equilibrium. It is worth noting that each parameter is determined independently and not "deduced" from a "closed loop" system condition.

3.4.9.1. *Parameters*

Total weight: 6.061 kg. Basic metabolism: $M_{et,b} =$ 2.4 W/kg. Respiratory flow rate: $\dot{V}_{resp} = 1.8.10^{-5}$ m^3/s.

h,sk	h,co	tr,skr	tr,m	tr,co	bl,cp	mat
279.4	3,243	739.5	6,883	3,804.6	2,009.7	2,340

Table 3.13. *Heat capacities (J/kg)*

h,sk-h,co	tr,sk-tr,m	tr,m-tr,co	h,sk-mat	tr,sk-mat	mat-ext
1.9	5.79	4	0.6	0.6	0.6

Table 3.14. *Conduction coefficients (W/K)*

Surface areas: head $A_h = 5.64.10^{-2}$m²; trunk $A_{tr} = 0.153$ m²; c_o = 0.75.

Transfer coefficients: $h_{cv} = 3.5$ W/m²K; $h_{ray} = 6$ W/m²K; F = 0.67; $k/h_{cv} = 10^{-3}$ K/Pa.

Textile filtering: $\lambda = 0.04$ W/mK; $i_v = 3.08$; $L_w = 10^3$ J/m^{3}K.

Ambient air: $T_a = 20$°C; $\varphi_a = 0.5$.

Blood flow $\dot{m}_{blood,b} = 4.5.10^{-5}$ kg/s; $c_{p,blood} = 4{,}187$ J/kgK.

We are going to test the influence of the clothing thickness parameter e_p, and this will subsequently determine $f_{cl} = A_v/A_{tr} = (D_{tr}+e_p)/D_{tr}$; $R_{th} = e_p/\lambda$; F_{cl} equation [3.53]; F_{pcl} equation [3.56].

As mentioned previously, thermal regulation is effected through vasodilatation and through transpiration starting from $T_n = 37{,}5$°C. The function $\omega(T_n)$ is considered linear starting from 0.05 at 37.5°C up to 0.5 at 39°C. Results obtained in the climatic chamber show that this temperature may be reached in case of sustained effort and subsequent abundant sweating. Likewise, the maximum value of skin moistening is generally estimated in the range (0.5–0.8). Finally, a metabolism corrective (f_{met}) [3.37] is considered for $T_n > 37$°C.

The simulation of equilibrium states obtained for a fixed thickness e_p provides the values of vasodilatation and transpiration parameters. Here are three equilibrium cases for a light clothing/blanket set.

e_p (mm)	$\dot{m}_{blood}/\dot{m}_{blood,b}$	ω	$T_{h,sk}$ (°C)	$T_{tr,sk}$	$T_{tr,co}$
6	1	0.05	33.2	34	36.2
7	1.8	0.05	33.5	34.7	36.9
10	20	0.084	34.7	35.8	37.5

Table 3.15. *Equilibrium cases for a light coverlet*

The first case creates the conditions for a mild cold sensation for which the model is not adapted. The second case shows thermal adaptation through a small increase in vasodilatation blood flow. The third case shows mild transpiration with maximum vasodilatation.

Should the clothing/blanket wrapping continue to increase, sweating will increase to the maximum. As the head's maximum evaporative capacity is reached, the "system" heats up, particularly because of the increase in metabolic flux (f_{met}). Figure 3.9 also shows this increase of the head's wet skin surface ratio.

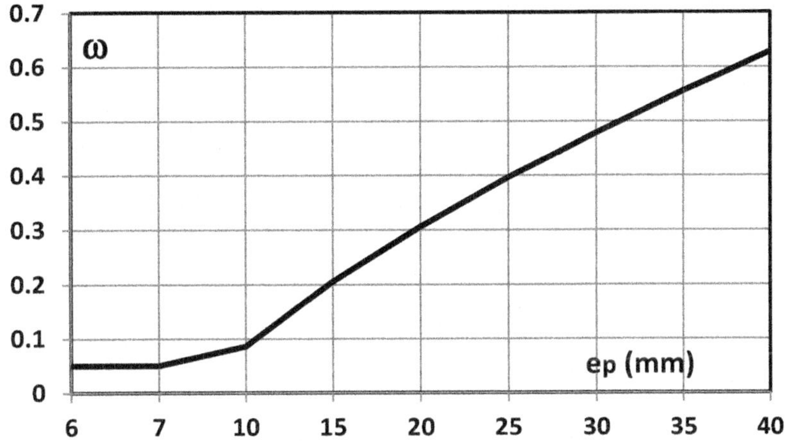

Figure 3.9. *Wet head skin surface area as a function of bedding thickness at thermal equilibrium*

The increase of this ratio is accompanied by intensified evaporation, which reduces head skin temperature (Figure 3.10), and by steady increase of central temperature.

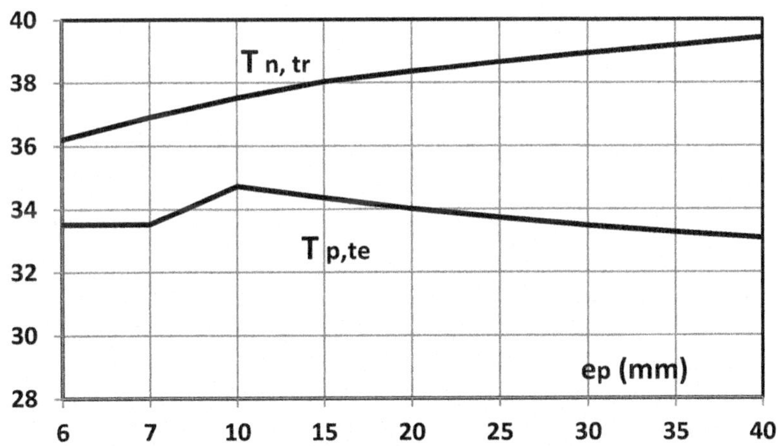

Figure 3.10. *Temperatures of head/skin ($T_{h,sk}$) and trunk/core ($T_{tr,co}$) at equilibrium as a function of bedding thickness*

The distribution of various heat fluxes dissipated by the baby in a comfort situation (e_p = 7 mm) and in a threshold overheating situation before heat-up (e_p = 40 mm) is shown in the following table and illustrated in Figure 3.11.

e_p (mm)	balance (W)	conv,h	ray,h	evap,h	sens,h	evap,tr	resp	mattress
7	14.55	13.7%	15.7	3.8	38.5	3.9	22.7	1.6
40	18.16	10.6%	12.2	36.5	10	11.6	17.5	1.4

Table 3.16. *Distribution of heat fluxes for a comfort case (ep=7 mm) and a overheated case (ep=40 mm)*

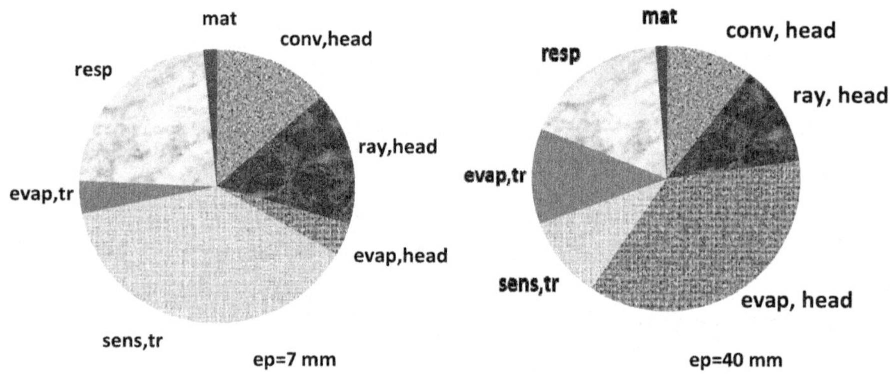

Figure 3.11. *Distribution of heat exchanges for a baby in two bedding situations*

These results support certain hypotheses advanced at the beginning of the modeling: the flux in the mattress is very secondary, while evaporation on the head is highly important. Advice concerning bedding is therefore justified: head radiation and respiration fluxes are considerable, so the surrounding environment should be carefully controlled. All heating surfaces whose radiation reaches the head should be avoided, the respiration space around the head should be cleared, and sleeping on the stomach should not be recommended.

What is the time needed for the "baby system" to reach such a state of thermal "equilibrium"? Is this some sort of time constant of the system? To answer these questions the system of equations should be solved starting from an initial state. For example, let us consider an initial state, when the baby is put to sleep, corresponding to a comfort state at e_p = 7 mm.

The blanket state is defined with a thickness $e_p = 40$ mm. Figure 3.12 shows the increase in temperature of the trunk core. After an hour, asymptotic increase is practically reached. A more precise definition of the system's "heat-up" time depends on the fixed maximum value of the wet head skin ratio (between 0.5 and 0.8), which cannot be measured directly. It is safe to consider that this time is less than 2 h.

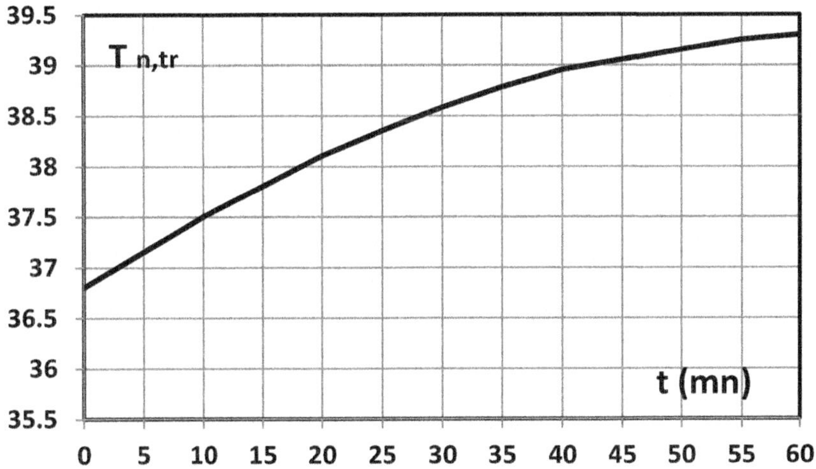

Figure 3.12. *Increase in temperature of the trunk-core (e_p = 40 mm)*

4

Heat and Humidity Transfer in Clothing

Since ancient times, clothing has mirrored the history of human development and has followed the evolution of techniques linked with scientific progress. For centuries, clothing based on the weaving of natural fibers such as cotton, wool, linen and silk has evolved as a result of the progress made in both mechanical weaving techniques and the production of natural dyes. The considerable advances in the chemical industry since the 19th Century have led to a large number of artificial fibers, natural wood-based polymers such as viscose and acetate and petroleum-based synthetic polymers such as polyester, polyamides and polyethylene being brought to market under a wide variety of brand names (Nylon®, Tergal®, Rhovyl®, etc.). Clothing from artificial fibers has become widespread in numerous other industrial domains, such as transport, industry and construction. Examples include: fiberglass used in naval construction, carbon fiber used in aviation and cycling and coated fabrics used in architecture.

Nowadays, the clothing sector is characterized by a wide distribution and extreme variety, as much in terms of usage (such as comfort, sporting uses and specialized work wear) as for the diversity of combinations of fibers and textile layers. Consider, for example, waterproof and "breathable" clothing, warm and lightweight fleeces and multi-layer anoraks. Studying the physics of heat and humidity transfer through such materials is a complex multi-parameter problem and providing a general mathematical model is a progressive, multi-stage approach.

4.1. From heterogeneous porous to continuous model media

Let us examine a fabric sample under a microscope (Figure 4.1). The interwoven yarn is made up of a bundle of twisted fibers. The physical properties derive from the nature of the fibers and from the type of weaving or knitting.

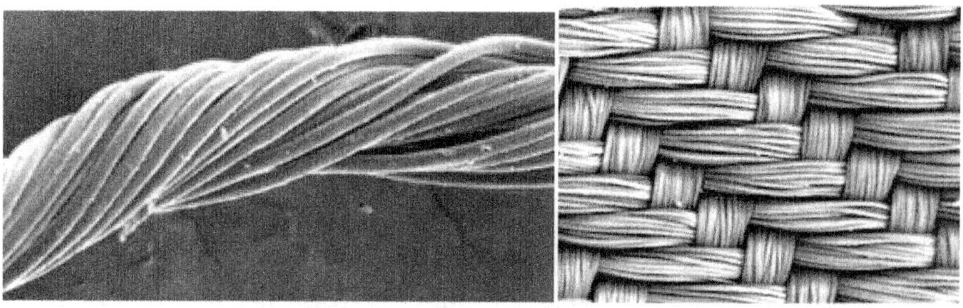

Figure 4.1. *Fiber in yarn and woven form*

Artificial fibers are only cylindrical as a first approximation, since being obtained through yarn drawing may result in varied surface states (aliasing) and may conceal a central hole (hollow fiber). Weaving and knitting provide quite diverse surface states, such as fabric inflation, piling and weaving which is either too tight or too loose. Cold protection fabric is designed to store a maximum of air in microscopic "cavities" generated by the fiber/yarn geometry (for example, clothes known as fleeces). Summer clothing should promote air circulation *via* "large holes" from weaving. In order of width, the dimension of common fiber is of the order of 10 to 20 µm and microfiber below 10 microns. For example, a 20 µm fiber results in a molded yarn of 50 fibers of around 1 mm and a fabric thickness of 2 mm. The "large holes", visible through transparency are of the order of the yarn diameter and the "small holes" which store air (or water) are of the order of the fiber diameter. To qualify fabrics which are particularly inflating is not so clear.

Fabric is therefore a very specific "porous" medium (to air, vapor and water) which can be considered as generally presenting a two peak distribution (Figure 4.2).

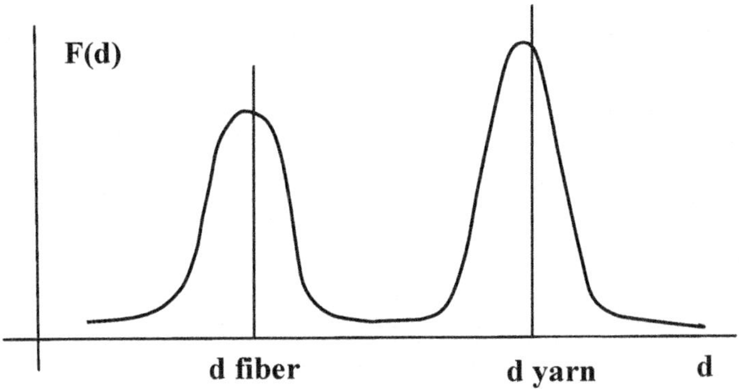

Figure 4.2. *Distribution of pores for a fabric sample*

The journey from scale and sampling of fabric to work clothing is a complex "system" approach, which involves many factors: 3D geometry, heterogeneities, circulation of air through interstices, heat transfers and complex radiation. One industrial approach is to test thermal transfers on a warmly clothed mannequin and to thus deduce comfort criteria in certain situations, such as transport and habitat. In this case we propose a detailed approach to studying transfers, but based on a single dimensional geometry relating to textile layer cross-sections. In other words, what heat and humidity transfers through a stack of diverse textile layers should be used for the varied representative limit conditions for worn clothing? The scientific approach is ahead of the industrial approach, which interprets scientific data based on experimental evidence.

The dress code for winter in particular is generally multilayer, where each layer is not necessarily homogeneous. Let us take for example a raincoat comprising an outer waterproof layer (resin or mousse, for instance) over a woven inner layer. This raises the problem of defining the homogeneous or "homogenizable" basic layer and this clothing can then be interpreted in principle as being comprised of two homogeneous layers. Moreover, the layers are wrapped around the body and may then present air gaps, according to surface states, which will act as what might be termed as contact resistance between physical layers. This equivalent layer of air, insofar as its thickness is small in comparison with the adjacent textile layers, will in principle be a secondary perturbation factor in the transfers by diffusion of heat and vapor, but will be able to be a primary factor in the "pumping" of liquid water (sweat).

Furthermore, the legitimacy of the operation to transform a particular "porous physical" textile layer into a "continuous mathematical" layer has been theoretically studied [WHI 98]. General conservation equations in continuous media will be of the same type, but with coefficients (such as conductivity and diffusivity) which take account of the integration of the original discontinuous medium. These coefficients will thus be the subject of specific sampling measurement in order to be used in mathematical modeling.

We now question the laws for the transfers of heat, vapor and liquid water (such as sweat and rain) for this textile medium.

4.2. Heat diffusion and convection

We must distinguish here two types of heat transfer: with or without forced filtration through the layer. Indeed everyone has the experience of wearing a pullover outside on a windy day; the air filtration finds its way through the textile layers in order to create a sensation of cold. The modern term "windbreak" defines clothing comprising a very tight outer layer, which impedes or limits this filtration. This filtration area-averaged velocity (V_f) can be measured from a suppression Δp between the sides "facing" and "against" the wind for the dressed person according to the law: $V_f = K\Delta p$, where K is the air permeability which can easily be measured in the laboratory, over a forced filtration device through one or more textile layers, based on air flow rate and the difference in pressure [standard ISO 9237]. Of what order of magnitude is this velocity? For example, take a tightly woven garment of $K = 10^{-3}$m/s Pa and a wind velocity of $V_{wind} = 10$ km/h, resulting in a suppression of $\Delta p = 1/2\rho_g V_{wind}^2$, $\rho_g = 1.2$ kg/m^3, from which we obtain: $V_f = 5.1$ mm/s. This velocity is often higher than the equivalent of vapor or liquid diffusion through textile layers.

A first formulation of the energy conservation equation, enthalpy balance, perpendicular to a physically homogeneous textile layer, is [WHI 98]:

$$\rho c_p \frac{\partial T}{\partial t} + \left(\rho c_p\right)_g V_f \frac{\partial T}{\partial x} = \frac{\partial}{\partial x}\left(\lambda_{eff} \frac{\partial T}{\partial x}\right) + \dot{q}_{source} \qquad [4.1]$$

Coefficients resulting from the operation of transition to the continuous medium should be stated. Indeed, considering a basic textile layer volume $d\mathcal{V}$ comprising of solid, gas (air and water vapor) and liquid water, the averaging solid and liquid densities are denoted by ρ_s, ρ_l, whereas the intrinsic densities of dry air and water vapor are denoted by ρ_a^i, ρ_v^i and relate to a volume fraction of gaseous space ε_g. The basic mass in the volume $d\mathcal{V}$ is written: $\rho\, d\mathcal{V} = (\rho_s + \rho_l + (\rho_a^i + \rho_v^i)\,\varepsilon_g)\, d\mathcal{V}$. The thermal capacity per unit of volume is thus written:

$$\rho c_p = \rho_s c_{ps} + \rho_l c_{pl} + (\rho_a^i c_{pa} + \rho_v^i c_{pv})\varepsilon_g \qquad [4.2]$$

The value g refers to the gaseous medium. The filtration velocity is measured (or calculated) from a flow rate across a "porous" basic surface and the gaseous convected mass flux is expressed as:

$$(\rho c_p)_g V_f = (\rho_a^i c_{pa} + \rho_v^i c_{pv}) V_f \qquad [4.3a]$$

Other mass transfers should be taken into account when considering the "convective" expression. Vapor is displaced by diffusion, especially across the large "holes" from weaving (velocity V_v) and liquid water is displaced as a result of capillary "pumping" (velocity V_l). The following should be added to the convective expression for forced filtration [4.3a]:

$$\rho_v^i c_{pv} V_v + \rho_l^i c_{pl} V_l \qquad [4.3b]$$

Note that these pseudo velocities are diffusion rates which we will express later based on bulk density gradients. The term $\dot{q}_{source}$ is the sum of all of the sources of internal "heat": condensation/evaporation and solar or infrared radiation. Due to the low variations in pressure and low velocities brought into play, we do not consider the effects of air compression and expansion (which are important in thermal machines) or viscous dissipation in this context. The energy conservation equation to be used is therefore unsteady and should express four heat transfer modes: conduction, convection, radiation and state change.

4.3. Vapor diffusion

Water vapor diffuses across the network of yarn/fiber interstices and a basic diffusion expression based on Fick's general law can be determined:

$$\rho_v^i V_v = -D_{eff}\rho_g^i \frac{d}{dx}\left(\frac{\rho_v^i}{\rho_g^i}\right) \qquad [4.4]$$

where $\rho_g^i = \rho_a^i + \rho_v^i$.

This flux can be expressed as a function of ρ_v^i and T based on the perfect gas characteristics of both dry air and vapor:

$$p_a = \rho_a^i r_a T;\ p_v = \rho_v^i r_v T;\ p_g = p_a + p_v \qquad [4.5]$$

Thus:

$$\rho_v^i V_v = -D_{eff}\left(\alpha_1 \frac{d\rho_v^i}{dx} + \alpha_2 \frac{dT}{dx}\right) \qquad [4.6]$$

where:

$$\alpha_1 = \left(1 + \rho_v^i \frac{T}{\alpha_0}\right)^{-1};\ \alpha_0 = \frac{p_g}{r_a - r_v};\ \alpha_2 = \left(\frac{T}{\rho_v^i} + \frac{T^2}{\alpha_0}\right)^{-1} \qquad [4.7]$$

Note that $\alpha_1 \cong 1$, $\alpha_2 << 1$, which can be used to justify the rejection of the term thermal diffusion and to then justify the use of the basic Fick's law. Insofar as calculations are programmed we can keep all of the secondary terms. The three coefficients α are calculated for average values of the variables. Pressure p_g is expressed as a constant value, which is a particular assumption for this singular porous environment. This may appear to be in contradiction with forced filtration, which assumes a total pressure gradient. Note that the expected variation in total pressure (and therefore the variation in α_0) within the textile layers has a negligible influence on coefficients α_1 and α_2. The mass conservation equation comprises one term for convection and another for diffusion.

How do we measure the heat and vapor diffusion coefficients?

Vapor and heat diffusion are both of the same type, but in this porous medium, cannot be deduced from each other. Coefficients λ_{eff} and D_{eff} are derived by specific measuring equipment. The most common device measures thermal or evaporative "resistance" from one or more sample layers of fabric brushed by an air current (Figure 4.3). This "skin model" device is regulated by a standard (ISO 11092).

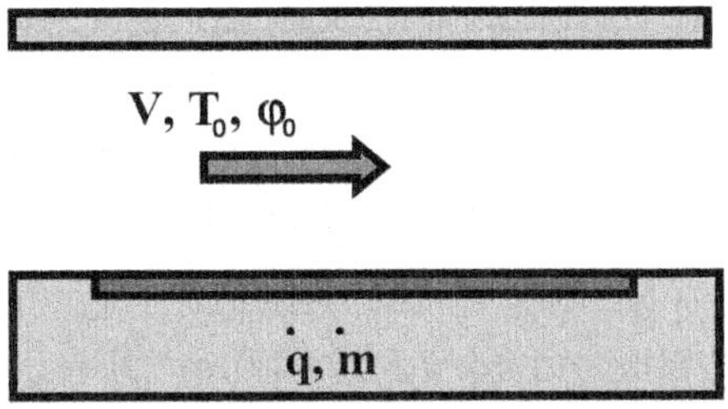

Figure 4.3. *Thermal and vapor conductivity measuring device*

The fabric sample of thickness e_p is placed on a plate warmed by electrical resistance which maintains the surface temperature T_p of the plate (but not the fabric) at 35°C. The electrical power per surface unit $\dot{q}$ is measured. The air current imposed is regulated in terms of velocity (1 m/s), temperature (20°C, in order to obtain thermal resistance and 35°C for evaporative resistance) and humidity (65%). In order to obtain evaporative resistance, the porous support plate is linked to a water supply system. A membrane between the plate and the material allows vapor to pass through, but not liquid water.

The summary results given by the standard are as follows. The thermal resistance of the textile layer and the boundary layer above in steady state is $R_{ct1} = \frac{T_p - T_o}{\dot{q}_1}$. The thermal resistance of the boundary layer without fabric is $R_{ct2} = \frac{T_p - T_o}{\dot{q}_2}$ and the thermal resistance of the sample is obtained by the difference $R_{ct} = R_{ct1} - R_{ct2} = \frac{e_p}{\lambda_{eff}}$. A more detailed analysis [BEN 90] shows the imprecision of the method for fine fabric. Indeed, the thermal resistance of the boundary layer is much higher than that of the fabric and the superficial exchange at the fabric's surface is not equal to that of the plain surface. We can assume that surface "coarseness" magnifies the exchange. The uncertainty over the thermal resistance may be 100% for fine fabric. Ben Aïm prescribes measuring multiple superposed layers. Moreover a corrective should be applied in order to take account of radiation which is not distributed in the same way, with or without the textile layer. For good orders of magnitude, Ben Aïm proposes to start from a compilation of results $R_{ct} = -0.0015+0.0248\ e_p$, where e_p is in mm and R_{ct} is in m^2K/W. We note that the thermal resistance of a textile layer is always lower than that of a still layer of air. Note also that the value λ_{eff} obtained is a global conductive coefficient which conceals a (minimal) participation of an exchange by high long wave radiation between the yarns and fibers [STU 84].

The results from evaporative resistance R_{et} derive from the same procedure as for thermal resistance. If R_{et1} refers to the boundary layer with fabric and R_{et2} refers to the boundary layer without, then $R_{et} = R_{et2}-R_{et1}$, but these are defined from the partial vapor pressure: $R_{et} = \frac{p_{vp}-p_{vo}}{\dot{q}}$. If we take into account the perfect gas equation [4.5], and from the fact that the electrical power is transformed during vapor flux $\dot{q} = \frac{D_{eff}}{e_p}\left(\rho^i_{vsat}(T_p) - \rho^i_{vo}\right)\Delta h_v$, this resistance is expressed:

$$R_{et} = \frac{p_{vp}-p_{vo}}{\dot{q}} = \frac{e_p}{D_{eff}}\frac{r_v T_p}{\Delta h_v} \qquad [4.8]$$

These measurement results bring into play a thickness uncertainty factor of e_p. Indeed the various preparations carried out by industry (such as brushing and inflation) indicate a heterogeneous material density which is compressed at the slightest contact. A conventional measurement of thickness can be expressed by applying a defined pressure, such as 1 kPa [standard ISO 5084].

These diffusion coefficients are a function of the moisture of the fabric. Indeed the porous structure of the medium is modified, even deformed in the presence of liquid water (swelling), which leads to modifications for heat conduction and vapor diffusion. The effect on thermal conduction is obvious: liquid water is a much better conductor than air of heat and conductivity strongly increases with moisture content (Figure 4.4) defined by $X = \rho_l/\rho_s$.

Examination of the curve indicates a particular behavior for low levels of moisture, which signifies sorption, or bound water, which is analyzed in the next section.

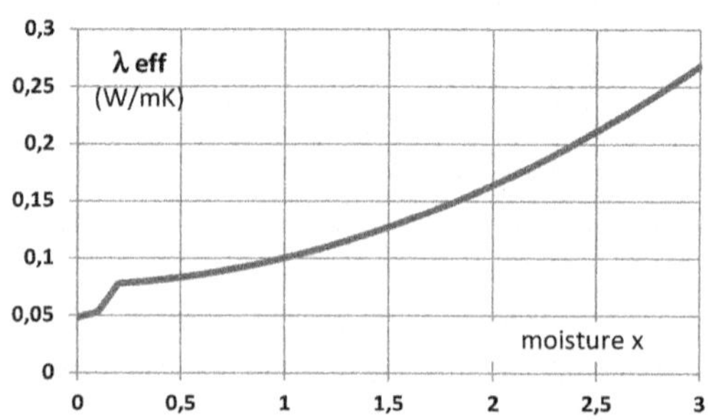

Figure 4.4. *Effective thermal conductivity as a function of moisture for cotton*

Vapor diffusivity is also dependent on moisture, but this dependence is of little practical interest. Indeed, while free liquid water is significant, displacement of liquid water from capillary forces is preponderant, and while water is only in bound form, the low layer at the fiber wall has little impact on the porous meshing which determines vapor diffusion. Vapor diffusivity D_{eff} depends on the "tightness" of the weaving and its order of magnitude is from 10^{-6} to $2*10^{-5}$ m²/s, and must be less than its value in air at rest ($2.63*10^{-5}$ m²/s at 25°C). From this data the order of magnitude of the diffusion rate is deduced. Take a typical textile layer used in clothing, of thickness 2 mm, which is subjected to an increase in temperature and humidity: skin side T = 34°C, $\varphi = 1$, outer side T = 20°C, $\varphi = 0.5$. Functions of humid air (Appendix 2) therefore give $\rho_v^i(skin) = 48.4*10^{-3}$ kg/m³; $\rho_v^i(out) = 9.7*10^{-3}$ kg/m³. Equation [4.6] with $\alpha_1 = 1$, $\alpha_2 = 0$, $D_{eff} = 10^{-6}$ to 10^{-5} m²/s, ρ_v^i as a factor of V_v, average of two inner and outer vapor densities, gives V_v = 0.67 to 6.7 mm/s, which signifies that the sole establishment of the diffusive vapor condition is very rapid.

Conductivity λ_{eff} of the order of 0.03 W/mK gives the order of magnitude of the establishment time scale for the heat diffusion condition. Indeed, equation [4.1] without filtration and without internal heat source shows that the diffusion time scale is of the order of $\tau \approx \frac{\rho c_p}{\lambda_{eff}} \Delta x^2$. From the following representative values: ρ = 300 kg/m³; c_p = 1,300 J/kgK; λ_{eff} = 0.03 W/mK; Δx = 1 mm, then τ = 13 s.

4.4. The effect of bound water

Cotton and wool are, like paper, highly hygroscopic fibers which are placed into equilibrium with ambient air humidity. Everyone has experienced the modification of these materials based on the season or local conditions. Leonardo da Vinci prescribed a method for determining air humidity by weighing a sample of cotton. Water which is bound or adsorbed is in the form of a layer several molecules in thickness, bound by electrical forces at the surface of the fibers. It can diffuse the length of the fibers with a specific coefficient. Its change of state heat (as a result of condensation/evaporation) is higher than that of free water. The maximum quantities of bound water (moisture) are therefore quite low, which is more significant for natural fibers (20% for cotton and 30% for wool) than for synthetic fibers (3% for polyester). This characteristic is widely emphasized by manufacturers regarding the layer in direct contact with the skin: cotton absorbs sweat better, but therefore takes longer to dry on skin.

Let ρ_b denote the bulk density of this bound water in a basic volume. From this point, we will denote ρ_l as the bulk density of the free liquid water which only exists when "saturation" of bound water has been achieved. A sample of yarn or textile is set at equilibrium in a controlled air humidity atmosphere (T, φ) according to a distribution (Figure 4.5):

$$\frac{\rho_b}{\rho_s} = f(\varphi); \quad \varphi = \frac{\rho_v^i}{\rho_{vsat}(T)} \qquad [4.9]$$

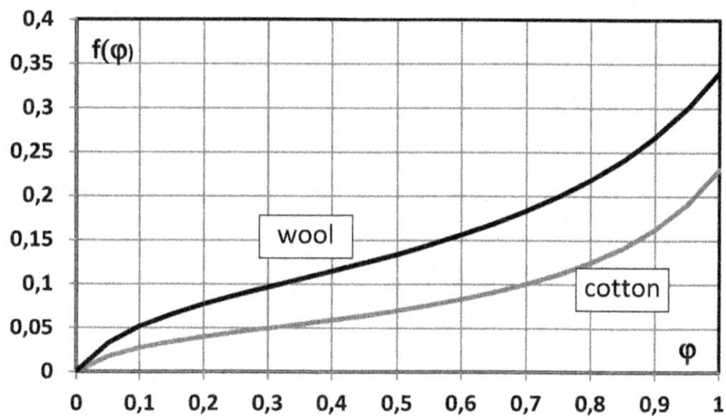

Figure 4.5. *Sorption equilibrium of cotton and wool*

Inside the textile medium during dynamic evolution, can we consider that humid air is in local equilibrium with the adsorbed water? To answer this, we must visualize the structure of the medium (fiber/yarn/weave) and follow the diffusion of the vapor in this interlacing. While the thickness of the adsorbed layer represents two or three molecules, adsorption from a fiber surface can be considered to be instantaneous. However, the "macroscopization" operation which transforms a heterogeneous porous medium into a fictive continuous one creates an effective vapor diffusion (D_{eff}) which "erases" the local effects. A delicate experiment may emphasize the distortion, in this context, between a physical continuous medium, such as a sheet of cellophane, and the continuous pseudo-medium of a textile layer [WEH 88]. One sorption cell enables the variation in the sample mass to be recorded and thus the quantity of adsorbed water and the crossing mass flux. The initially dry sample is subjected to increases in water vapor without any increase in temperature. The sheet of cellophane reacts as a homogeneous medium (Figure 4.6): there is an adsorption phase in the medium, followed by a displacement flux phase (see problem of the wall subjected to an increase in temperature). When the vapor gradient is applied, the textile layer displays an increasing vapor flux which accompanies the internal adsorption (Figure 4.7). The explanation is clear: the "large holes" from weaving facilitate a diffusion flux which only slightly releases water on the "polished" fibers. Adsorption is slowed by vapor diffusion between fibers of the yarn and probably diffusion of bound water along the fibers or across the fiber wall when it is hollow (as with wool). Mathematical formalization of this behavior gives rise to many possible expressions, while introducing coefficients which are tricky to obtain. The simplest model consists of expressing that equilibrium between humid air and adsorbed water is locally satisfied while expressing a possible equivalent diffusion from this water in the fictive homogeneous medium of the form:

$$\rho_b V_b = -D_b \frac{\partial \rho_b}{\partial x}; \ \frac{\rho_b}{\rho_s} = f(\varphi) \qquad [4.10]$$

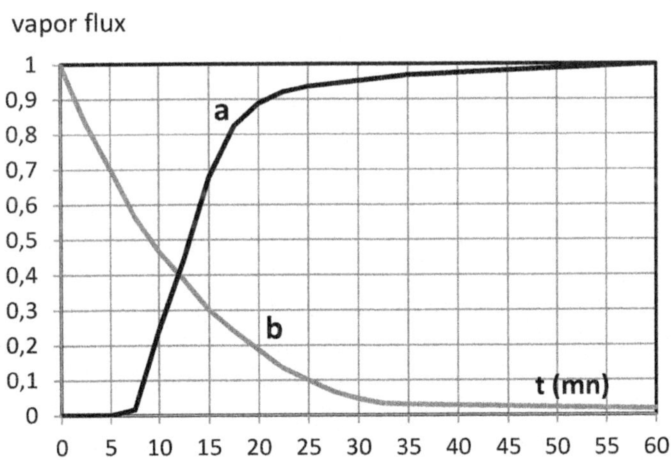

Figure 4.6. *Sheet of cellophane subjected to a vapor gradient: a) crossing flux, b) adsorbed flux. Fluxes reported to their maximum values*

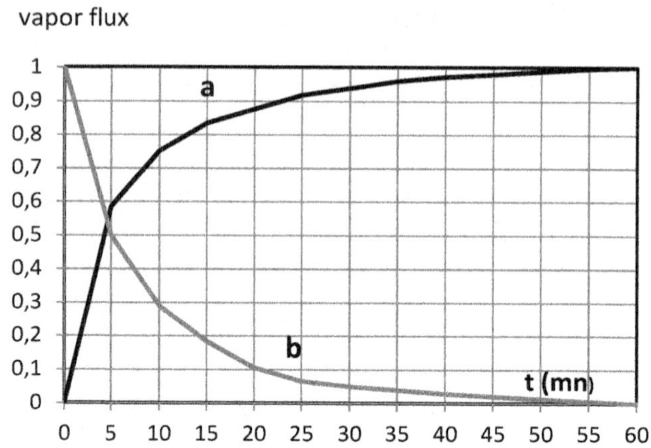

Figure 4.7. *Sample of textile (cotton) subjected to a vapor gradient: a) crossing flux, b) adsorbed flux*

This adsorption model, when applied to a textile layer subjected to a vapor gradient, provides results which are closer to the sheet of cellophane than for the textile layer [FOH 02]. When applied to cloth, the approximation introduced is generally insignificant. Remember that the quantities of adsorbed water are very low compared with those of free water, from sweat for instance.

In order to model the behavior of highly hygroscopic natural fiber textiles (such as cotton and wool), it may be important to have good orders of magnitude for this diffusion coefficient D_b, and the question is how to determine this coefficient. The principle is to imagine a protocol which emphasizes this behavior while "eliminating" all of the others, i.e. free water, air infiltration, temperature gradient. This is the principle behind Wehner's "manipulation" [WEH 88] and may equally be the case at the end of the controlled drying of a textile layer when all of the free water has been evacuated and temperature may be considered to be constant. The dynamic equations which govern the loss (or gain) in bound water are the mass conservation equations for vapor and bound water:

$$\varepsilon_v \frac{\partial \rho_v^i}{\partial t} + \frac{\partial \rho_v^i V_v}{\partial x} = -\dot{m}_b \qquad [4.11]$$

$$\frac{\partial \rho_b}{\partial t} + \frac{\partial \rho_b V_b}{\partial x} = \dot{m}_b \qquad [4.12]$$

where ε_v is the volume proportion of vacuum available for humid air and $\dot{m}_b$ is the volume flux of condensation from bound water vapor (negative for evaporation).

The bound water/vapor local equilibrium model indicates a "parallel" bound water/vapor diffusion with exchange of water and to these equations [4.11] and [4.12] it is necessary to add [4.6], [4.9] which is transformed at constant temperature into:

$$\partial \rho_b = \alpha_3 \partial \rho_v^i;\ \alpha_3 = \frac{\rho_s}{\rho_{vsat}(T)} \frac{df}{d\varphi} \quad [4.13]$$

Adding [4.11] and [4.12] then gives:

$$\left(\varepsilon_g + \alpha_3\right) \frac{\partial \rho_v^i}{\partial t} = \frac{\partial}{\partial x}\left(\left(D_{eff}\alpha_1 + D_b\alpha_3\right) \frac{\partial \rho_v^i}{\partial x}\right) \quad [4.14]$$

This equation is of the classic diffusive type when $D_{eq} = D_{eff}\alpha_1 + D_b\alpha_3$ and noting that $\varepsilon_v \ll \alpha_3$. It is necessary to add a boundary condition to the edges of the textile layer. The two examples quoted above are reduced to:

$$-D_{eq}\left(\frac{\partial \rho_v^i}{\partial x}\right)_{x=e_p} = k\left(\rho_{vp}^i - \rho_{vo}^i\right);\ -D_{eq}\left(\frac{\partial \rho_v^i}{\partial x}\right)_{x=0} = 0 \quad [4.15]$$

e_p is the thickness of the layer (or half-layer, according to the example). k is the mass exchange coefficient determined by the nature of the boundary layer for which the vapor density is denoted by ρ_{vp}^i at the textile layer wall and ρ_{vo}^i at the exterior. Equilibrium with humid air at the end of drying is expressed:

$$\frac{\rho_b}{\rho_s} = f(\varphi);\ \varphi = \frac{\rho_{vo}^i}{\rho_{vsat}(T)}.$$

Since there are therefore results available based on evidence providing the evolution of moisture over time $X(t) = \frac{1}{\rho_s e_p} \int_0^{e_p} \left(\rho_b + \varepsilon_g \rho_v^i\right) dx$, how do we deduce D_{eq}?

Integration over x of equation [4.14], taking [4.15] into account, provides:

$$\rho_s e_p \frac{dX}{dt} = \left(D_{eq} \frac{\partial \rho_v^i}{\partial x}\right)_p = -k\left(\rho_{vp}^i - \rho_{v\infty}^i\right) \quad [4.16]$$

In order to solve this, we introduce a parabolic humidity profile which is generally true for "slow" drying. So, in order to satisfy [4.15]: $\rho_b = a(t)x^2 + b(t)$, which leads to:

$$X = \frac{1}{\rho_s e_p}\left(a(t)\frac{e_p^3}{3} + b(t)e_p\right) \quad [4.17]$$

[4.16] becomes:

$$\rho_s e_p \frac{dX}{dt} = 2ae_p\left(\frac{D_{eq}}{\alpha_3}\right)_p = -k\left(\rho_v^i(\rho_{bp}) - \rho_{vo}^i\right) \quad [4.18]$$

By linearizing [4.13] the vapor gap becomes a bound water gap:

$$-k\left(\rho_v^i(\rho_{bp}) - \rho_{vo}^i\right) = -\frac{k}{\alpha_3}\left(\rho_{bp} - \rho_{bo}\right) \quad [4.19]$$

Equation [4.18] thus gives:

$$\rho_s e_p \frac{dX}{dt} = -\frac{k}{\alpha_3}\left(ae_p^2 + b - \rho_{bo}\right) \quad [4.20]$$

which, coupled with [4.17], constitutes a system of two equations with two unknowns, a and b.

Therefore: $a = -\frac{3\rho_s}{2e_p^2}\left(X + \frac{e_p\alpha_3}{k}\frac{dX}{dt} - f(\varphi)\right)$ and the equation for D_{eq} drawn from [4.18] is expressed:

$$D_{eq}(\rho_{bp}) = -\frac{\alpha_3 e_p^2}{3}\frac{\frac{dX}{dt}}{\left(X + \frac{e_p\alpha_3}{k}\frac{dX}{dt} - f(\varphi)\right)} \quad [4.21]$$

This expression is often simplified when we are able to introduce $\frac{e_p\alpha_3}{k}\frac{dX}{dt} \ll X$ and the following form is commonly used:

$$\frac{dX}{dt} = -D_{eq2}(X - X_o);\ D_{eq2} = D_{eq}\frac{3}{\alpha_3 e_p^2};\ X_o = f(\varphi) \quad [4.22]$$

Let us take an example of determination of D_b based on an experiment to dry a cotton fabric sample, deposited on the plate of a scale (e_p = 1mm, ρ_s = 330 kg/m^3). A small air current is passed over the textile layer (T = 20°C, φ = 0.5, k = 0.015 m/s). The sorption curve for this type of cotton is defined by $f = \varphi/(c_1+c_2\varphi+c_3\varphi^2)$, c_1 = 1.985, c_2 = 18.29, c_3 = 15.93; therefore X_o = 0.07.

The calculations provide $\rho_{vsat}(20°C) = 0.017$ kg/m^3, $\frac{\partial f}{\partial \varphi} \approx 0.2$, $\alpha_3 \approx 3,800$, $e\rho\alpha_3/k \approx 260$ s. The order of magnitude of the measured values is introduced: $X \approx 0.1$, $dX/dt \approx 2*10^{-4}$ s^{-1}, and [4.21] gives $D_{eq} \approx 2.6*10^{-6}$ m^2/s, then with a value $D_{eff} = 2*10^{-6}$ m^2/s, $D_b \approx 1.5*10^{-10}$ m^2/s. Precision of the value of D_b therefore depends on the precision of related parameters k, e_p, $f(\varphi)$ and D_{eff}. The adsorption/desorption behavior of bound water associated with humid air has greater time constants than those from vapor or heat diffusion. From the numeric values above and [4.14], a time scale can be deduced: $\tau \approx \alpha_3 \frac{e_p^2}{D_{eq}} = 24$ min.

In order to consider the impact of condensation and evaporation it should be noted that bound water "sticks" more strongly to the solid wall than the so-called "free" liquid water and therefore a stronger energy is required for the change of state. The following formulation illustrates this [MOR 93]:

$$\Delta h_b = \Delta h_v \left(1 + C_0 exp\left(-C_1 \frac{\rho_b}{\rho_s}\right)\right) \qquad [4.23]$$

The following indicative values are used: cotton $C_0 = 0.5135$, $C_1 = 26.976$; wool $C_0 = 0.565$, $C_1 = 11.23$; polyester $C_0 = 0.435$, $C_1 = 38.47$.

4.5. Liquid water diffusion

Liquid water in clothing may be caused by sweat, rain or from vapor condensation. It is physically imposed onto the bound water (adsorbed). A flux of displaced liquid water firstly fills the "bound water reservoir" before diffusing into the porous network. Similarly, a condensing vapor flux saturates the material with bound water before producing "free" water from the movement resulting from the forces of capillarity and eventually gravity.

The effects of capillary forces, or of surface tension, are used in everyday life to "mop" or "wipe" liquid surfaces (for example using towels or hydrophilic cotton) or to fill a dropper tank. Water rising from capillary pressure p_c in a thin 1/10 mm tube known as a capillary tube is a well-known behavior and Jurin's law expresses the height h attained:

$$p_c = p - p_a = \frac{2\sigma}{r} cos\theta = \rho g h \qquad [4.24]$$

where σ is the superficial tension of the liquid, r is the radius of the tube, ρ is the density of water and θ is the contact angle at the wall.

On very clean glass, θ is virtually zero and the free surface is a half sphere. Stains (grease molecules) on the glass modify the contact angle and therefore the wetting characteristic. Everyone knows that mercury does not wet glass ($\theta<0$). The wetting characteristics of cotton are generally very high, those of wool and synthetic fibers such as polyester very low. This characteristic is occasionally exploited in order to obtain "water repellant" surfaces. The following experiment is easy to carry out with a sample of this type of textile – tightly woven, often microfiber, but without coating of a layer which is opaque to movement of air and water. For this kind of textile, the advert speaks about breathable waterproof clothing. An open tube is closed at one end by this fabric sample and water is gradually poured into the other end. When the water level reaches height h the water filters through the fabric (Figure 4.8).

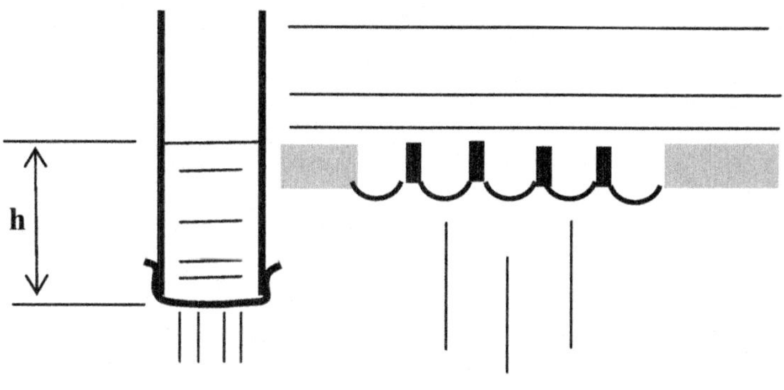

Figure 4.8. *Water is forced through water-breathable waterproof textile*

Water is released when the force due to pressure from gravity is greater than the surface tension, i.e. $\rho_l gh>2\sigma/r$ if $\theta = 0$ during discharge, where r is the radius of a typical pore from the weaving (inter-yarn). We state the orders of magnitude: $r = 10^{-2}$ cm, $\rho = 1$ g/cm^3, $\sigma = 75$ cgs units and $g = 981$ cm/s^2 gives us $h = 15.2$ cm.

This behavior explains how waterproof membranes, which are introduced between two textile layers, work: the air passage gaps are small and the water drop slides off (but does not spread over) the surface ($\theta<0$). The argument according to which the gaps allow air molecules but not “over large” water molecules to pass through is purely commercial.

Liquid water introduced at a point will diffuse into the layer and from one layer to another through capillary “pumping”. The purposes is to introduce in principle a diffusion equation which is similar to a heat equation by assuming that it is possible to determine experimentally the diffusion coefficient. Measurement methods analyze the liquid diffusion in the textile layer plane and it must be assumed that the resulting diffusion coefficient is of the same order through the thickness of the textile layer.

Let us consider the wetting of absorbent textiles such as cotton and the following basic experiment: a 20 cm strip of cotton, sliced parallel to the warp or weft, is suspended above a bowl of water, just touching the liquid surface (Figure 4.9).

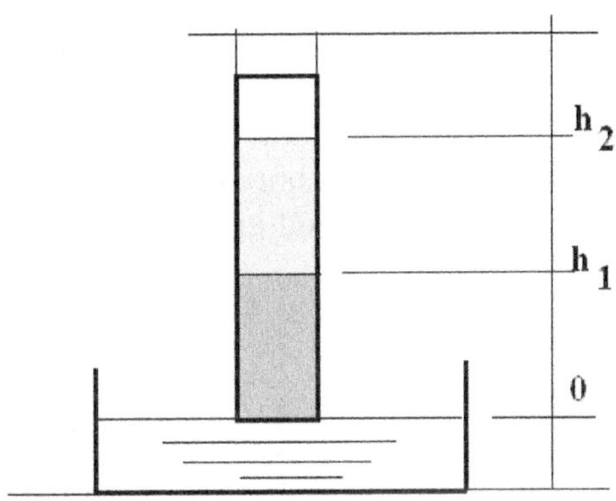

Figure 4.9. *Liquid absorption onto a band of cotton*

A watchful eye, possibly helped by using a magnifying glass, or better still a microscopic viewfinder, shows two areas of quite limited wetting density of heights $h_1$and h_2.

If the cotton strip is not too narrow, i.e. contains enough water, then it is possible to “read” its distribution over a paper towel. It is sufficient to arrange the cotton strip over an unfolded paper towel and placed on a non-absorbent horizontal surface. Pat it lightly with a finger to ensure uniform wetting and observe the wet area of the paper. The residual deformation of the paper still provides an outline of the wetting area (Figure 4.10). This indicates that as a first approximation the wetting density is distributed into two virtually uniform areas, corresponding to inter-fiber and inter-yarn porosity.

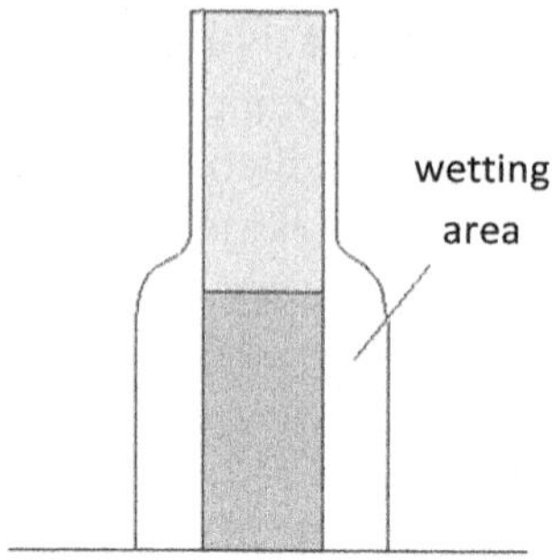

Figure 4.10. *Water absorption onto a sheet of paper*

Before considering the general behavior as a result of a diffusion coefficient, we present an explanatory model of flow in cylindrical pores to help the readers.

Consider capillary pumping on a horizontal fabric strip (parallel to the warp or weft). One end (x = 0) is dipped in water at t = 0.

The quantity of movement equation for the flow in a pore of radius r_0 (which is constant) is:

$$\rho\frac{\partial u}{\partial t}+\rho u\frac{\partial u}{\partial x}=-\frac{\partial p}{\partial x}+\mu(\frac{\partial^2 u}{\partial r^2}+\frac{1}{r}\frac{\partial u}{\partial r}) \qquad [4.25]$$

Due to mass conservation (ρ constant), u(r,t).

Integrate over the radius, noting the averaged velocity U:

$$\rho\frac{dU}{dt}=-\frac{\partial p}{\partial x}-8\mu\frac{U}{r_0^2} \qquad [4.26]$$

Integration of the pressure over distance x between x = 0 ($p = p_a$) and $x = x_0$ ($p = p_a - p_c$) gives:

$$\left(\rho\frac{dU}{dt}+8\mu\frac{U}{r_0^2}\right)x_0=p_c=2\frac{\sigma}{r_0}\cos\theta \qquad [4.27]$$

The first term on the left hand side of the equation is the inertial term, the second is the viscous resistance term; on the right hand side we have the motor term (capillary pressure). The order of magnitude of the inertial to viscous term ratio is $\frac{\rho r_0^2}{8\mu t}$. t is a value for time characteristic of "pumping", of the order of several minutes. When $r_0 = 10^{-2}$cm, $\rho = 1$g/cm^3, $\mu = 9*10^{-3}$ g/cms, t = 60 s we arrive at a ratio of $2*10^{-5}$. This means that the quasi-static hypothesis is found.

Equation [4.27] therefore gives:

$$U = \frac{dx_0}{dt} = \frac{\sigma r_0}{4\mu x_0}\cos\theta \qquad [4.28a]$$

which is integrated into:

$$x_0 = \sqrt{\frac{\sigma\cos\theta}{2\mu} r_0 t} \qquad [4.28b]$$

The displacement of the liquid diffusion "front" is the square root of r_0t. The value for x_0 using $r_0^{1/2}$ may puzzle the readers. Indeed, the capillarity experiment with thin vertical tubes plunged into a water-filled receptacle indicates that the height of the water level is inversely proportional to the tube diameter. Here we are referring to an imbibition experiment, where, viewing through magnifying glasses, we clearly see small inter-fiber "pores" filled before "large" interyarn pores, which appears to contradict the cylindrical model leading to [4.28b]. A detailed examination using a magnifying glass reveals inter-fiber capillary displacement and displacement along the yarn grooves; then, when the weft and warp yarn is sufficiently filled, the capillary film "jumps" in order to fill the holes from the textile "mesh". The large "holes" from weaving do not lead to the formation of a physical channel enabling the capillary increase as with yarn. We therefore see a first "imbibition front" before the second, which saturates the medium.

Is it possible to quantify this behavior using a diffusion equation and how can the corresponding coefficient be experimentally determined?

Let the unidirectional diffusion equation be as follows:

$$\frac{\partial X}{\partial t} = \frac{\partial}{\partial x}\left(D_l \frac{\partial X}{\partial x}\right); X = \frac{\rho_l}{\rho_s} \qquad [4.29]$$

and let the imbibition limit conditions from a horizontal strip (parallel to warp and weft) be:

$$X(x,t); X(0,t) = X_{max}; \ \forall t, x \to \infty, X \to X_{ini} \quad [4.30]$$

With D_l constant, equation [4.29] enables changes in the variable x,t in a unique variable $s = \frac{x}{2\sqrt{D_l t}}$, which is consistent with [4.28], and assumes a simple solution [CRA 75, BEJ 93] for an analogous heat diffusion problem:

$$\frac{X - X_{ini}}{X_{max} - X_{ini}} = 1 - \frac{2}{\sqrt{\pi}} \int_0^s exp(-u^2)du \quad [4.31]$$

The readers will recognize the error function (Figure 4.11), $X_+ = (X - X_{ini})/(X_{max} - X_{ini})$.

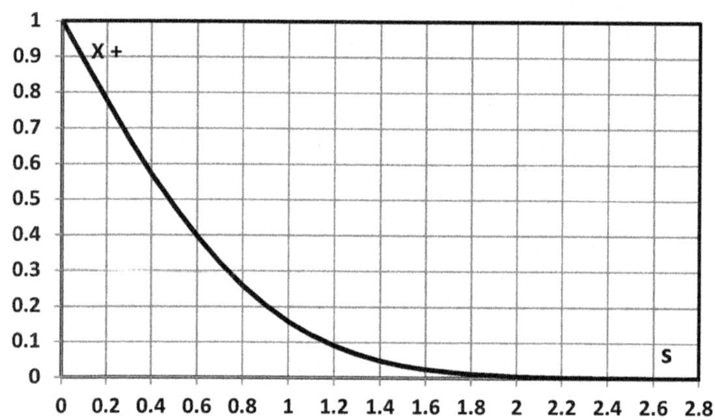

Figure 4.11. *Imbibition profile for constant D_l $X+=(X-X_{ini})/((X_{max}-X_{ini})$*

This solution infers a method for determining D_l. A sliced strip of fabric is placed on a measuring device cutting off parallel to the welt or warp and then on a flat non-absorbent surface. The end of the strip is plunged into the water at time t = 0. A number of detection methods for the movement of the imbibition front exist. For example: conductivity meter, or optical or infrared camera (to measure the temperature difference between dry and wet points). We assume that we can note the value of the front passage for all values constant of X_+ e.g. 0.5, therefore for all constant values of s, let $s = s_{ind}$. From time t_1 to t_2 the front has progressed from x_1 to x_2, so if $s_{ind} = \frac{x_1}{2\sqrt{D_l t_1}} = \frac{x_2}{2\sqrt{D_l t_2}}$ then:

$$x_2 - x_1 = 2 s_{ind} \sqrt{D_l} (\sqrt{t_2} - \sqrt{t_1}) \quad [4.32]$$

The value of D_l deduced from [4.32] can be determined from a number of pairs of points (x_1, x_2), which enables the pertinence of the model to be estimated. Coefficient D_l is slightly different parallel to warp or weft and depends heavily on the nature of the fibers and the weaving. Microfibers diffuse liquid more quickly and are therefore beneficial in clothing in supplying water from sweat to the external evaporation surface. Equation [4.31], however, shows what is proven through experimentation, the existence of an imbibition "front" which spreads over time. Indeed, if we trace functions $X_+(x)$ over time t_1 and time $t_2 > t_1$, then $X_{+2}(x) > X_{+1}(x)$.

Other wetting devices are possible, such as from a point (e.g. a cotton wick or a dripper) on a textile surface resting on a table. We observe the development of an elliptical humidity spot which visualizes the two principal warp and weft directions. The wetted area can be recorded using, for instance, an infrared camera. The solution for the cylindrical coordinate diffusion equation in a homogeneous medium may be used for processing purposes. It is then sufficient to compare the theoretical wetted circle at the surface of the ellipse in order to obtain a "mean" diffusion coefficient between the warp and weft directions. This equation is written:

$$\frac{\partial X}{\partial t} = \frac{1}{r} \frac{\partial}{\partial r} \left(r D_l \frac{\partial X}{\partial r} \right); X(r,t) \quad [4.33]$$

With the following limit conditions:

$$X(0,t) = X_{max};\ \forall t, r \to \infty, X \to X_{ini}$$

At constant D_l this problem is similar to the thermal problem of heat diffusion from a continuous source line [BEJ 93 p. 181], but with a condition causing a continuous source (a constant mass flow rate). The analytical solution is obtained by converting variables r, t into a single variable $s = \frac{r}{\sqrt{D_l t}}$:

$$\frac{\partial^2 X}{\partial s^2} = -\left(\frac{s}{2} + \frac{1}{s} \right) \frac{\partial X}{\partial s} \quad [4.34]$$

For which the solution is:

$$X - X_{ini} = \frac{\dot{q}}{4\pi D_l} \int_{\frac{s^2}{4}}^{\infty} \frac{1}{u} exp(-u)du \qquad [4.35]$$

Using the definition of X, the normalized mass flow rate at the source is written: $\dot{q} = \frac{\rho_l}{\rho_s}\dot{q}_v$, where $\dot{q}_v$ is the effective volume flow rate. If the wetting front is detected for a fixed value of X and therefore of s, the same reasoning as before can be applied to obtain a wetted area S:

$$S = \pi r^2 = \pi s^2 D_l t \qquad [4.36]$$

The wetted area is a linear function of time, and experimentation proves this perfectly for an elliptical area. A coefficient D_l can be deduced from this. The rate of displacement of the wetted surface from the measurement of a area difference ΔS during a time interval Δt gives $D_l = \frac{\Delta S}{\Delta t}\frac{1}{\pi s^2}$. Measuring a delivered liquid flow rate (drop by drop, for example) gives: $\frac{4\pi D_l}{\dot{q}}(X - X_{ini}) = F(s)$, where F(s) is the integral of [4.35]. It follows from this that:

$$\frac{\Delta S}{\Delta t}\frac{X - X_{ini}}{\dot{q}} = \frac{s^2}{4} \int_{\frac{s^2}{4}}^{\infty} \frac{1}{u} exp(-u)du \qquad [4.37]$$

A value for s and therefore D_l correspond to a threshold detection X.

With the objective being to obtain a good amount of transversal liquid diffusion, for low thicknesses, therefore, before the front spreads, large values of X_+ are indicated.

Values for the liquid diffusion coefficient D_l range from 10^{-7} to 10^{-6} m²/s, which provides values for liquid diffusion time scale through a layer 1 mm thick: $\tau \approx \frac{e_p^2}{D_l}$ from 1 to 10 s.

What is the liquid water diffusion law for vertical fabric? Equations [4.29] and [4.33] are used to derive the following from the balance (without condensation or evaporation):

$$\frac{\partial \rho_l}{\partial t} + \vec{\nabla}.\left(\rho_l^i \vec{V}_l\right) = 0; \ \rho_l^i \vec{V}_l = -D_l \vec{\nabla} \rho_l \qquad [4.38]$$

Gravitational force is added to an expression which is more general than [4.38] according to Darcy's law:

$$\vec{V_l} = -\frac{K}{\mu_l}\left(\vec{\nabla} p_l - \rho_l^i \vec{g}\right) \quad [4.39]$$

Where K is the permeability of the medium during water flow and p_l is the liquid pressure linked to the capillary pressure by $p_l = p_g - p_c(\rho_l)$. Geologists and Hydraulic engineers often use this law in order to predict underground flows. For textile layers we generally assume a constant value p_g (local atmospheric pressure) and the expression for unidirectional conservation equation [4.38] combined with [4.39] becomes:

$$\rho_l^i \vec{V_l} = -D_l \frac{\partial \rho_l}{\partial x}\vec{x} + \frac{K\rho_l^i}{\mu_l}\rho_l^i \vec{g};\ D_l = -\frac{K}{\mu_l}\rho_l^i \frac{dp_c}{d\rho_l} \quad [4.40]$$

Capillary pressure may also depend on temperature, which might introduce a supplementary term which we will be ignoring here as we are provided with the limited range of temperature differences. For textiles, it is more complex to find distribution $p_c(\rho_l)$ than to determine D_l using a particular experiment. The expression for the term for gravity can also be used under certain circumstances.

4.6. Mass and energy balances

The mass conservation equation at a point of a textile layer is a balance between bound water, liquid water and vapor, taking account of state changes such as condensation and evaporation. Each type has its own balance equation; for bound water (equations [4.10] and [4.12]):

$$\frac{\partial \rho_b}{\partial t} + \frac{\partial \rho_b V_b}{\partial x} = \dot{m}_b;\ \rho_b V_b = -D_b \frac{\partial \rho_b}{\partial x};\ \frac{\rho_b}{\rho_s} = f(\varphi) \quad [4.41]$$

For liquid water (without a gravity term):

$$\frac{\partial \rho_l}{\partial t} + \frac{\partial \rho_l^i v_l}{\partial x} = \dot{m}_l;\ \rho_l V_l = -D_l \frac{\partial \rho_l}{\partial x} \quad [4.42]$$

Note the difference between ρ_l, the liquid density (mass of water divided by the basic volume around the point), and ρ_l^i, the intrinsic density of water.

The velocity V_l is therefore a flow rate velocity in the direction x. $\dot{m}_l > 0$ for condensation and < 0 for evaporation.

For vapor (see equations [4.6] and [4.11]):

$$\varepsilon_g \frac{\partial \rho_v^i}{\partial t} + \frac{\partial \rho_v^i V_v}{\partial x} + V_f \frac{\partial \rho_v^i}{\partial x} = -\dot{m}_b - \dot{m}_l \quad [4.43]$$

$$\rho_v^i V_v = -D_{eff}\left(\alpha_1 \frac{\partial \rho_v^i}{\partial x} + \alpha_2 \frac{\partial T}{\partial x}\right) \quad [4.44]$$

Conservation of humidity from vapor, bound water and free water is obtained through the addition of equations [4.41]–[4.43]:

$$\frac{\partial}{\partial t}\left(\varepsilon_g \rho_v^i + \rho_b + \rho_l\right) = -V_f \frac{\partial \rho_v^i}{\partial x} + \frac{\partial}{\partial x}\left(D_{eff}\left(\alpha_1 \frac{\partial \rho_v^i}{\partial x} + \alpha_2 \frac{\partial T}{\partial x}\right) + D_b \frac{\partial \rho_b}{\partial x} + D_l \frac{\partial \rho_l}{\partial x}\right) \quad [4.45]$$

The gradient of ρ_b can be expressed as a function of the vapor and temperature gradients, based on the local equilibrium hypothesis [4.9]:

$$\frac{\partial \rho_b}{\partial x} = \alpha_3 \frac{\partial \rho_v^i}{\partial x} - \alpha_4 \frac{\partial T}{\partial x};\ \alpha_3 = \frac{\rho_s}{\rho_{vsat}(T)} \frac{df}{d\varphi};\ \alpha_4 = \alpha_3 \varphi \frac{d\rho_{vsat}}{dT} \quad [4.46]$$

The time derivative of ρ_b can be justified using the same formalization. Equation [4.45] (without a filtration term) becomes:

$$\left(\varepsilon_g + \alpha_3\right)\frac{\partial \rho_v^i}{\partial t} - \alpha_4 \frac{\partial T}{\partial t} + \frac{\partial \rho_l}{\partial t} = \frac{\partial}{\partial x} \left(\left(D_{eff}\alpha_1 + D_b \alpha_3\right)\frac{\partial \rho_v^i}{\partial x} + \left(D_{eff}\alpha_2 - D_b \alpha_4\right)\frac{\partial T}{\partial x} + D_l \frac{\partial \rho_l}{\partial x}\right) \quad [4.47]$$

Energy equation [4.1] is completed using the expression for sources of heat from state change and radiation and the convective/diffusive heat fluxes (forced filtration, vapor, bound water and liquid water). The flow of bound water is completely negligible.

$$\rho c_p \frac{\partial T}{\partial t} + \left(\sum \rho c_p V\right)\frac{\partial T}{\partial x} = \frac{\partial}{\partial x}\left(\lambda_{eff} \frac{\partial T}{\partial x}\right) + \Delta h_{vb} \dot{m}_b + \Delta h_v \dot{m}_l + \dot{q}_{ray} \quad [4.48a]$$

$$\left(\sum \rho c_p V\right) = \left(\rho c_p\right)_g V_f + \rho_v^i c_{pv} V_v + \rho_l^i c_{pl} V_l \quad [4.48b]$$

The term for the thermal flux of bound water is neglected. The term for the vapor thermal flux will often be negligible compared with the conductive flux.

The flow rates of condensation and evaporation are extracted from [4.41], [4.42] and so the preceding equation becomes:

$$\left(\frac{\rho c_p}{\Delta h_v} + \alpha_4 r_h\right)\frac{\partial T}{\partial t} - \alpha_3 r_h \frac{\partial \rho_v^i}{\partial t} - \frac{\partial \rho_l}{\partial t} = -\frac{(\Sigma \rho c_p V)}{\Delta h_v}\frac{\partial T}{\partial x} + \frac{\partial}{\partial x}\left(\frac{\lambda_{eff}}{\Delta h_v}\frac{\partial T}{\partial x} - D_l \frac{\partial \rho_l}{\partial x}\right) - r_h \frac{\partial}{\partial x}\left(D_b\left(\alpha_3 \frac{\partial \rho_v^i}{\partial x} - \alpha_4 \frac{\partial T}{\partial x}\right)\right) + \frac{\dot{q}_{ray}}{\Delta h_v} \quad [4.49]$$

where $r_h = \Delta h_{vb}/\Delta h_v$ and the state change enthalpy Δh_v being taken to be constant within the temperature range considered.

These fairly general equations are simplified based on the context. In particular, the filtration imposed can cancel out the heat, bound water and vapor diffusion behaviors and certainly disturbs the liquid water diffusion. The coefficients introduced may be functions of x and t through a physical parameter such as $\lambda_{eff}(\rho_l)$ or $\alpha(T)$.

The two balance equations obtained link the physical variables ρ_v^i, ρ_l and T based on x and t. In order to close the system we must add the physical criteria of condensation and evaporation:

$$\rho_v^i < \rho_{vsat}(T) \rightarrow \rho_l = 0; \; \rho_l > 0 \rightarrow \rho_v^i = \rho_{vsat}(T) \quad [4.50]$$

4.7. Limit conditions

The "system" considered consists of one or more textile layers, a physical layer resulting from manufacturing or a layer decomposed in order to satisfy the "homogenizable" criterion. Although the transfers may be not relative to a garment (3D) but transverse to the layers (1D), we refer to skin side (internal) and external conditions.

The skin side of thermal and hydric constraints may be debated. In Chapter 3 we saw possible expressions for convective [3.41], radiative [3.45] and evaporative [3.42] heat flux using the Sherwood and Nusselt number correlation in order to obtain the exchange coefficients for heat h_i and mass k_i. The problem is a little more complex here as we also need to take into account the presence of a liquid flow (sweat and rain). The skin side conditions will therefore need to be fixed to be most "realistic" following the intended simulation. For example, if we think about a layer of air, of a defined thickness, between the skin and the textile layers, then we can fix a skin temperature T_{skin} and a wetted skin fraction ω which enables the convected vapor and heat fluxes per unit of area to be expressed:

$$q_{cv,int} = h_i(T_{skin} - T_1);\ \dot{m}_{vap,int} = \omega k_i(\rho^i_{vsat}(T_{skin}) - \rho^i_{v1}) \quad [4.51]$$

The index 1 is relative to the point at the internal surface of the first layer. The contribution of the radiation flux to be considered (long wave or solar) will be discussed later.

In order to take account of a liquid flow (sweat) directly skin side of the textile layer we can use either a liquid saturation condition at point 1 ($\rho_l = \rho_{lmax}$), or a liquid flow $\dot{m}_{liq}$, which is compatible with the possibilities for textile layer pumping.

On the external side, if we take the index n for the external side of the latter layer, T_{ext} to be the external temperature and $\rho^i_{v,ext}$ to be the external vapor density, the heat and mass convected fluxes are:

$$q_{cv,ext} = h_e(T_n - T_{ext});\ \dot{m}_{vap,ext} = k_e(\rho^i_{vn} - \rho^i_{vext}) \quad [4.52]$$

The filtration velocity V_f is determined from a wind velocity V (or movement speed of a typical person), which generates excess pressure ½ $\rho_a V^2.C_p$ (see flow around the cylinder), where C_p is the pressure coefficient based on position on the obstacle. In this context we can be satisfied with the representative values for the cylinder: into the wind (C_p = 1) and downwind (C_p = -2). If filtration is based on n clothing layers for which we know the permeability K_i (defined in [4.2]) of each layer, the filtration rate, as a result of adding the pseudo-resistances, is:

$$V_f = -C_p \frac{1}{2} \rho_a V^2 \left[\sum_i \frac{1}{K_i}\right]^{-1} \quad [4.53]$$

The minus sign in [4.53] results from the fact that the x-axis crosses the multilayer between the skin and the outside.

A standard [ISO 9237] describes the protocol for obtaining the K coefficients from the measurement of an air flow rate across a textile layer over a pressure difference of 100 Pa. Suppose, for example, we have a multilayer composed of three layers: skin side a loosely woven layer of $K = 2*10^{-2}$ m/sPa, followed by a slightly tighter fabric layer of $K = 2*10^{-3}$ m/sPa and then a very tightly woven layer of $K = 7*10^{-4}$ m/sPa. The K coefficient which is equivalent to these three layers is then $K = 5.05*10^{-4}$ m/sPa, which, for a wind velocity of 10 km/h ($C_p = 1$), gives a filtration velocity $V_f = -2.3$ mm/s and for a wind velocity of 50 km/h, $V_f = -58$ mm/s.

For a human body, the problem is a little more complicated. The flow of air which penetrates through multiple layers facing the wind must exit either through openings (collar, sleeves) or by flowing a second time through the layers down wind, which will slightly modify the velocity calculated above.

Although the "coexistence" of filtration velocity and heat and vapor diffusion may be theoretically possible, the context of application according to the garment does not generally justify the deepening of this option, even if the different terms figure in the equations. We will see later how the threshold value for the filtration rate to be considered can be characterized.

If the composition comprises multiple physical layers (such as for winter clothing), examination of the constituent products and how they are combined may lead to either perfect contact between or a layer of air equivalent to an assumed "contact resistance", with this air being a break in the continuity of the liquid water.

4.8. Processing for a numerical resolution

The general numerical resolution of these equations is quite complex and the advantage of using a specific programming language for this purpose is that physical specificities, such as bound water and free water, can be better defined. The use of "multipurpose", but closed code means that minimal adaptation is required to the problem being addressed. It is possible to retain a system comprising physical variables from the start $(\rho_b, \rho_v^i, \rho_l, T, \dot{m}_b, \dot{m})$, the corresponding equations and the inequalities from exclusion.

For each simulation, it is strongly advised that equations should be solved without dimensions for which the principal terms are of the order of 1. The best reference values should be chosen for this operation. Some values are imposed, such as a reference thickness e_p, which may be multilayer thickness, temperature gap ΔT, such as the difference between skin and external temperature, vapor density $\rho^{i}_{vref} = \rho_{sat}(T_{skin})$, reference density of bound water ρ_{bref}, such as that from saturation for the most absorbent bound water layer, and so on. The reference time τ_{ref} should be selected from the most significant diffusion time scales for the problem. Indeed, we have seen previously that a textile layer of 1 mm thickness of cotton assumes characteristic time of the order: 10 s for heat diffusion, less than 1 s for vapor diffusion of bound water and 1 to 10 s for liquid water. Other characteristic times can also be introduced with boundary conditions conditions. The priority being to promote a physical phenomenon.

Therefore:

$$\rho_v^+ = \frac{\rho_v^i}{\rho_{vref}^i};\ \rho_b^+ = \frac{\rho_b}{\rho_{bref}};\ \rho_l^+ = \frac{\rho_l}{\rho_{lref}};\ T^+ = \frac{T - T_{ini}}{\Delta T_{ref}};\ x^+ = \frac{x}{e_p} \quad [4.54]$$

4.9. First example: condensation in a multilayer

Let us consider the problem of condensation in a piece of winter dress, which is examined for its insulating properties, but which is prone to condensation. This effect is particularly noticeable during physical activity, causing sweat in particular areas of the body. We are restricting ourselves here to investigating vapor skin exchanges and will simplify the mathematical model by suggesting some reasonable additional hypotheses at the start of condensation for current textiles: no forced filtration, no solar radiation, the effect of infrared radiation is taken into account through the value of coefficients, negligible bound water, steady vapor and thermal states, negligible liquid diffusion, $\alpha_1 = 1$, $\alpha_2 = 0$, $\alpha_3 = 0$, $\alpha_4 = 0$. The low heat and vapor diffusion time scales before condensation justifies the fall in the corresponding unsteady terms. The system of equations is thus considerably simplified into:

$$\frac{\partial}{\partial x}\left(D_{eff}\frac{\partial \rho_v^i}{\partial x}\right) = \frac{\partial \rho_l}{\partial t};\ \frac{\partial}{\partial x}\left(\lambda_{eff}\frac{\partial T}{\partial x}\right) = -\Delta h_v \frac{\partial \rho_l}{\partial t} \quad [4.55]$$

with existence conditions [4.50] and limit conditions [4.51], [4.52].

It is possible to combine this system, by eliminating ρ_l, as follows:

$$\begin{cases} \rho_v^i < \rho_{vsat}(T) \Rightarrow -\lambda_{eff}\dfrac{\partial T}{\partial x} = \varphi_T; -D_{eff}\dfrac{\partial \rho_v^i}{\partial x} = \varphi_\rho \\ \rho_v^i = \rho_{vsat}(T) \Rightarrow -\lambda_{eff}\dfrac{\partial T}{\partial x} - \Delta h_v D_{eff}\dfrac{\partial \rho_v^i}{\partial x} = \varphi_{energ} \end{cases} \quad [4.56]$$

The readers will recognize, on the second line of [4.56], the term for enthalpic energy conservation.

Let us assume the following reference values: $\rho_{vref}^i = \rho_{vsat}(T_{skin})$; $\Delta T_{ref} = T_{skin} - T_{ext}$. The reference for liquid water and time are not set here. Furthermore, let us take a reference value for thermal conductivity and vapor diffusivity: λ_{ref} , D_{ref}. The system above is written as follows, for numerical resolution using $u_\lambda = \dfrac{\lambda_{eff}}{\lambda_{ref}}$; $u_D = \dfrac{D_{eff}}{D_{ref}}$:

$$\begin{cases} \rho_v^i < \rho_{vsat}(T) \Rightarrow -u_\lambda\dfrac{\partial T^+}{\partial x^+} = \varphi_{T+}; -u_D\dfrac{\partial \rho_v^{i+}}{\partial x^+} = \varphi_{\rho+} \\ \rho_v^i = \rho_{vsat}(T) \Rightarrow -u_\lambda\dfrac{\partial T^+}{\partial x^+} - N u_D\dfrac{\partial \rho_v^{i+}}{\partial x^+} = \varphi_{energ+} \end{cases} \quad [4.57]$$

where: $\varphi_{T+} = \varphi_T \dfrac{e_p}{\Delta T_{ref}\lambda_{ref}}$; $\varphi_{\rho+} = \varphi_\rho \dfrac{e_p}{\rho_{vref}^i D_{ref}}$; $N = \dfrac{\Delta h_v D_{ref}\rho_{vref}^i}{\Delta T_{ref}\lambda_{ref}}$ a number without dimension.

Let us set the data from this multilayer comprising a tee-shirt on skin, a "fleece" pullover and a double-layer anorak. The question is about the intermediary air layers which exist in clothing which has not had too much in the way of adjustment. For this specific example, it is not suggested to introduce them, but to "integrate" them into the adjacent textile layer by modifying its thickness and eventually its physical coefficients. Indeed, condensation in an air layer equates to a "mist" which goes into clothing and settles on adjacent textile layers. The data is therefore as shown in Table 4.1.

	Tee-shirt	Fleece	Lining	Anorak
Thickness (mm)	2	3	1	2
λ_{eff} (W/mK)	0.06	0.04	0.04	0.04
D_{eff} (m^2/s)	$2*10^{-5}$	$2*10^{-5}$	$2*10^{-5}$	$8*10^{-6}$
u_λ	1.5	1	1	1
u_D	1	1	1	0.4

Table 4.1. *Multilayer data*

Skin side T_{skin} = 34°c, φ = 90%, h_{int} = 10 W/m²K, k_{int} = $5*10^{-3}$ m/s; external side T_{ext} = 0°C, φ = 50%, h_{ext} = 20 W/m²K, k_{ext} = 10^{-2} m/s. We deduce e_p = 8 mm, ρ^i_{vint} = 0.0337 kg/m³; ρ^i_{vext} = 0.0024 kg/m³. We use λ_{ref} = 0.04 W/mK, D_{ref} = $2*10^{-5}$ m²/s.

The last system can be solved by performing a nodal method to conserve fluxes. The multilayer is "cut" into 8 slices of thickness Δx = 1 mm and so $\Delta x_+ = \Delta x/e_p$ = 1/8 with 10 nodes at the center of each slice (i = 2 to 9) and 2 supplementary nodes on each free side (i = 1 skin side and i = 10 on the external side). With each volume around the node being at constant temperature and humidity, the boundary at the midpoint between nodes determines the distances necessary to calculate the resistances. The fluxes are therefore written as follows:

For i = 2 to 9 :

$$\varphi^+_{Ti,i+1} = \frac{T_i^+ - T_{i+1}^+}{R_{Ti,i+1}};\ R_{Ti,i+1} = \frac{\Delta x_+}{2}\left(\frac{1}{u_{Ti}} + \frac{1}{u_{Ti+1}}\right) \qquad [4.58]$$

For the skin side thermal flux:

$$\frac{h_{int} e_p}{\lambda_{ref}}\left(T^+_{skin} - T_1^+\right) = \frac{T_1^+ - T_2^+}{R_{T1,2}};\ R_{T1,2} = \frac{\Delta x_+}{2}\frac{1}{u_{T1}} \qquad [4.59]$$

The expression for the external side flux is thus easily written as well as the equivalent for mass fluxes.

Identifying a condensation area is easy. It is sufficient to resolve the first line of [4.57] and to plot $T^+(x)$, $\rho^{i+}_v(x)$ in order to observe that over a wide range $\rho_v^i > \rho_{vsat}(T)$. However, it is not possible to deduce from this that this is the effective condensation range since "heat from condensation" redistributes the temperatures. We must therefore determine nodes $P_1(x_1)$ and $P_2(x_2)$, $x_1<x_2$ at the start and end of condensation.

The energy conservation for the nodes of the condensation area (equation [4.57]) is written as follows:

$$\frac{T_i^+ - T_{i+1}^+}{R_{Ti,i+1}} + N\frac{\rho_{vi}^{i+} - \rho_{vi+1}^{i+}}{R_{\rho i,i+1}} = ;\varphi^+_{energ}\,;\ \rho_v^i = \rho_{vsat}(T);\ i = p_1, p_2 - 1 \qquad [4.60]$$

For the skin side non-condensed area (if it exists):

$$\frac{T^+_{skin}-T^+_{p1}}{\frac{1}{h^+_{int}}+\sum_1^{p_1} R_{Ti,i+1}} = \varphi^+_{Tint} \; ; \; h^+_{int} = \frac{h_{int}e_p}{\lambda_{ref}} \; ; \; \frac{\rho^+_{v,skin}-\rho^+_{v,p1}}{\frac{1}{k^+_i}+\sum_1^{p_1} R_{\rho i,i+1}} = \varphi^+_{\rho int} \; ; \; k^+_{int} = \frac{k_{int}e_p}{D_{ref}} \quad [4.61]$$

For the external side non-condensed area:

$$\frac{T^+_{p2}-T^+_{ext}}{\frac{1}{h^+_{ext}}+\sum_{p_2}^{n} R_{Ti,i+1}} = \varphi^+_{Text} \; ; \; \frac{\rho^+_{v,p2}-\rho^+_{vext}}{\frac{1}{k^+_{ext}}+\sum_1^{p_1} R_{\rho i,i+1}} = \varphi^+_{\rho ext} \quad [4.62]$$

Furthermore, the energy conservation at points P_1 and P_2 gives:

$$\varphi^+_{Tint} + N\varphi^+_{\rho int} = \varphi^+_{energ} = \varphi^+_{Text} + N\varphi^+_{\rho ext} \quad [4.63]$$

The number of equations [4.60]–[4.63] corresponds to the number of unknowns if we are given x_1 and x_2. An iteration is necessary in order to verify $\rho^+_{v,1} < \rho_{vsat}(T^+_i)$; $i < p_1$; $i > p_2$.

The following figures provide profiles for temperature and humidity in the multilayer, with or without condensation which is produced from nodes 2 to 8.

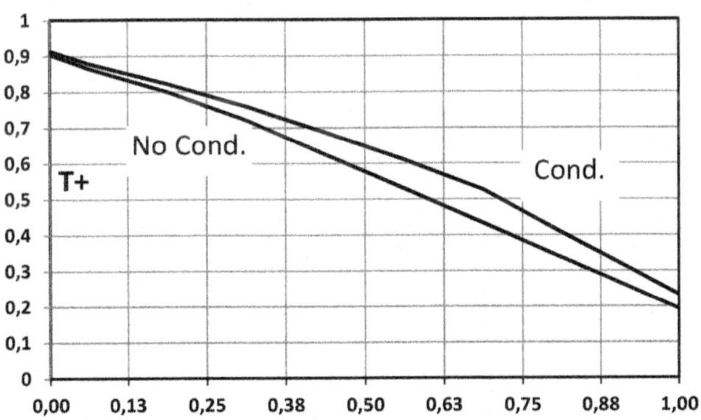

Figure 4.12. *Profile for temperature in the multilayer with (Cond) or without (No Cond) condensation*

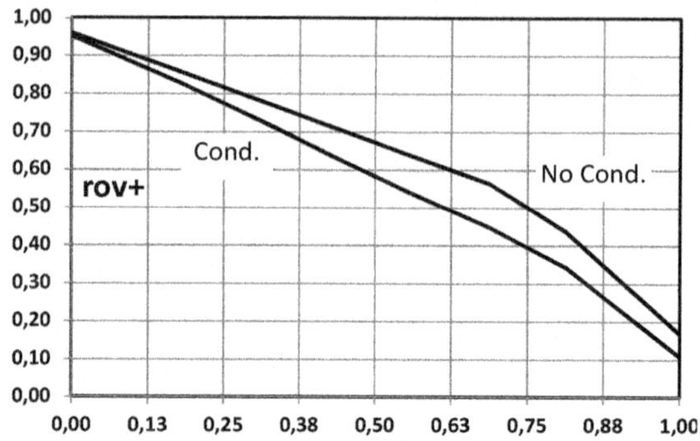

Figure 4.13. *Humidity profiles (ρ_{v+}) in the multilayer*

The condensed liquid flow rate in the multilayer can be deduced from these profiles. Indeed, integrating equation [4.55] gives:

$$\frac{\partial}{\partial t}\int_{x_1}^{x_2} \rho_l \, dx = \left[D_{eff}\frac{\partial \rho_v^i}{\partial x}\right]_{x_1}^{x_2} = \frac{D_{ref}\rho_{vref}}{e_p}\left[u_D\frac{\partial \rho_v^{i+}}{\partial x^+}\right]_{x_1}^{x_2} \qquad [4.64]$$

For the example above we obtain a condensation flow rate of 57 g/m² hour.

4.10. Convection and diffusion

The filtration velocity introduced above is characteristic of forced convection which may compete against the various diffusion behaviors (heat and liquid vapor) present in a multilayer. What is the value for the filtration velocity which generates these behaviors?

Consider the simple case of a heat transfer through convection and diffusion in a multilayer without condensation or internal heat source and in steady state. Equation [4.48] is simplified into:

$$\left(\rho c_p\right)_g V_f \frac{\partial T}{\partial x} = \frac{\partial}{\partial x}\left(\lambda_{eff}\frac{\partial T}{\partial x}\right) \qquad [4.65]$$

The filtration velocity V_f may be either positive or negative. Indeed, taking in the background the reference for a clothed individual subjected to wind and the x-axis through an area of the multilayer, moving from skin side outward, then, into the wind $V_f<0$ and downwind $V_f>0$.

Let us take limit conditions which are a little simpler than the previous example, i.e. $T(x = 0) = T_p$; $T(x = e_p) = T_{ext}$. The previous example indicates that these conditions marginally modify the expected result. For a homogeneous multilayer, equation [4.65] assumes an analytical solution:

$$\frac{T-T_{ext}}{T_{skin}-T_{ext}} = 1 - \frac{exp\left(\frac{Px}{e_p}\right)-1}{\exp(P)-1} \; ; P = \frac{(\rho c_p)_g V_f e_p}{\lambda_{eff}} \qquad [4.66]$$

P is the Peclet number which represents the ratio of convection to diffusion "forces". The equilibrium value P = 1 for these two behaviors will provide us with a representative filtration rate. If we have the following representative values: $(\rho c_p)_g$ = 1,200 J/m^{3}K; $e_p = 8*10^{-3}$m; λ_{eff} = 0.04 W/mK, then for P = 1, V_f = 4.2 mm/s.

Figure 4.14 shows the deformation in the temperature profile when the Peclet number is either positive or negative. Without filtration the profile is linear. This graphic should be compared with real life experience: without any leakage from the collar or sleeves, areas facing the wind (P<0) are more susceptible to cooling than those downwind (P>0) and this is in spite of the fact that downwind depression may be double the suppression into the wind (see Figure 2.7: pressure around a cylinder).

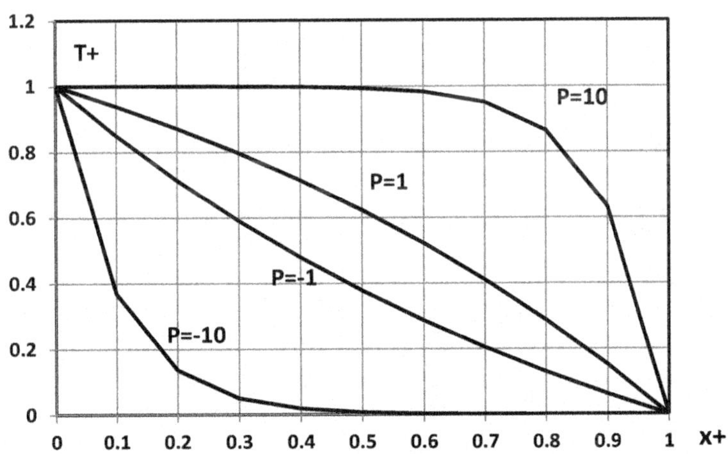

Figure 4.14. *Normalized temperature profiles for different Peclet values*

Let us now find an order of magnitude for heat flux $\dot{q}$ evacuated with a Peclet number P = -10. The filtration rate is therefore based on the preceding numeric values, $V_f = 41.7$ mm/s. From equation [4.66] if $\Delta T = T_{skin}-T_{ext}$:

$$\dot{q} = -\lambda \left(\frac{dT}{dx}\right)_{x=0} = \lambda \frac{\Delta T}{e_p} \frac{P-1}{\exp(P)-1} \qquad [4.67]$$

If $T_{skin} = 34°C$, $T_{ext} = 20°C$, then $\dot{q} = 770$ W.

Take the example of an individual on a motorcycle, riding at 100 km/h, i.e. V = 27.8 m/s, wearing a multilayer of tightly woven clothes of thickness e_p = 8 mm, with coefficient K = $5*10^{-4}$ m/sPa and conductivity λ = 0.04 W/mK. For the area "into the wind" (ρ_a = 1.2 kg/m$^{3;}$$c_{pg}$ = 10^3 J/kgK, C_p = 1), then: V_f = 23.2 cm/s, P = 55. This order of magnitude demonstrates the need for better suited clothing (such as leather and impermeable surfaces) in order for this filtration to be diminished.

The convection/diffusion problem introduces difficulties into numerical resolution, which becomes tricky and demands a suitable discretization scheme [PAT 80, Chapter 5].

4.11. Taking account of radiation

Radiation is a proven effect based on experience in terms of comfort either in the sunshine, which then necessitates suitable clothing, or in the home (heated flooring, chilled ceilings, fireplaces and wood stoves) and occasionally for certain professions (such as firefighters). It is therefore necessary to consider the two characteristics of radiation: "short wave" solar radiation and "long wave" radiation of walls. The coefficients which characterize materials for each type of radiation are, indeed, different.

A textile layer is a highly heterogeneous medium for radiation, which is redistributed through complex local geometry; we also measure the global properties of a layer in order to determine transmission (τ coefficient), absorption (α coefficient), and reflection (ρ coefficient).

The textiles industry, from clothing to fashion, plays on the assortment of fibers, the mechanical treatment of surfaces and the chemical treatment of coloration, in order to obtain "ad hoc" (occasionally based on wavelength –

this is the Moire effect), reflection coefficients. In order to take account of thermal effects, we will use some coefficients here, which are integrated over the wavelength range (solar or infrared) considered. Even taking the nature of the surface of the material into account, radiation is diffuse and non-directional, except clearly for beam solar radiation.

Generally, we have the following: $\alpha+\rho+\tau=1$ and for infrared flux Kirchoff's law for grey, diffuse surfaces provides equality for the emission and absorption coefficients: $\varepsilon = \alpha$.

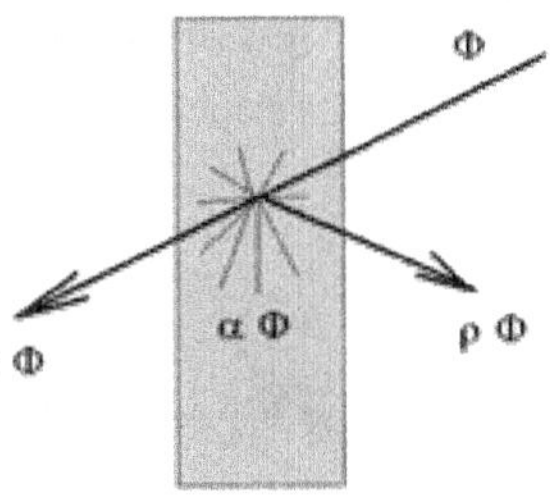

Figure 4.15. *Radiation on a textile layer*

Some measured values for solar radiation are shown below [see SAR 94, p. 48]:

Fabric type	Thickness mm	ρ_{sol}	τ_{sol}	α_{sol}
Tergal pants	0.315	0.39	0.37	0.24
Cotton joggers	1.77	0.37	0.25	0.38
Cotton tee-shirt	0.363	0.56	0.36	0.08
Cotton jeans	0.858	0.56	0.14	0.30

Table 4.2. *Solar radiation data for some textile samples*

For the same textile samples above, the following table provides measured values in an infrared environment:

Fabric type	Thickness mm	ρ_{IR}	τ_{IR}	α_{IR}
Tergal pants	0.315	0.27	0.18	0.55
Cotton joggers	1.77	0.31	0.27	0.42
Cotton tee-shirt	0.363	0.10	0.41	0.49
Cotton jeans	0.858	0.27	0.15	0.58

Table 4.3. *Infrared radiation data for the preceding samples*

Note that reflectivity is a partly volume and non-surface related behavior, as H. Ben Aïm [BEN 90] shows when measuring the reflection coefficients of multiple layers of the same fabric. For modern textile layers of the order of 0.5 mm thickness, the ρ coefficient may increase from 10 to 50% when we superimpose two or three layers and not increase thereafter. In order to consider layer fractions from the same textile, and to attribute coefficients to them based on those of the initial layer, it is convenient to follow a ray and its decomposition into two (or three) identical layers and to sum the reflection and transmission outputs. Let us take the example of two identical layers with parameters ρ, τ and α and search for parameters ρ_2, τ_2, and α_2 for the double layer with a radiation incident of value 1 (Figure 4.16). The radiation which, for ease of representation, appears to be directional, is actually generally diffuse.

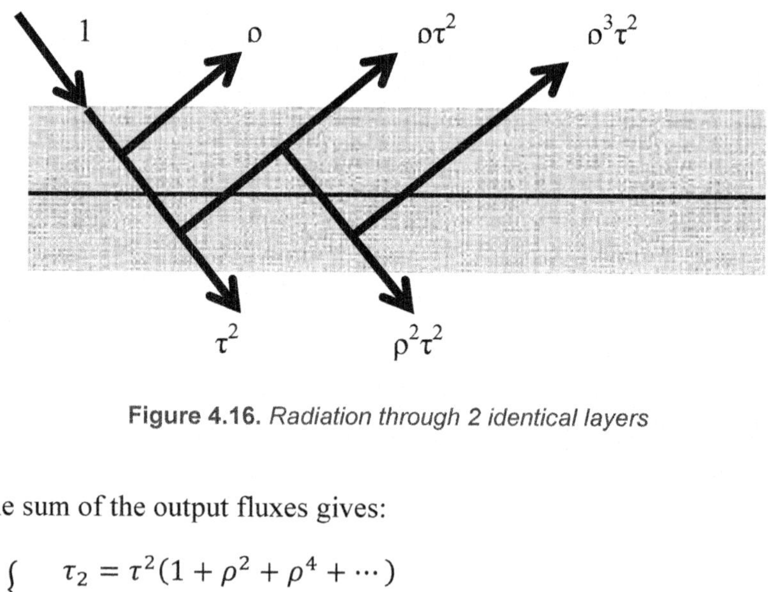

Figure 4.16. *Radiation through 2 identical layers*

The sum of the output fluxes gives:

$$\begin{cases} \tau_2 = \tau^2(1 + \rho^2 + \rho^4 + \cdots) \\ \rho_2 = \rho[1 + \tau^2(1 + \rho^2 + \rho^4 + \cdots)] \end{cases}$$

Note the development of the series:

$$\tau_2 = \frac{\tau^2}{1-\rho^2}; \; \rho_2 = \rho(1 + \tau_2) \qquad [4.68]$$

Inversely, from ρ_2, τ_2, we can deduce the pairing ρ, τ. Coefficient α is deduced from $\alpha = 1-\rho-\tau$.

Let us now write the balance equations for radiation considering a stationary or virtually stationary state. The significance of variations in the ρ,τ, and α coefficients, which is linked to modification of the substrate such as, for example, the appearance of liquid water, could be an interesting research topic.

Take n textile layers (or partial layers) which are coupled and for which external faces are subjected to solar radiation ϕ_{sol}. Additionally, we have at our disposal tools to calculate this based on meteorological conditions (Appendix 3). Consider three adjacent layers i-1, i and i+1 (Figure 4.17).

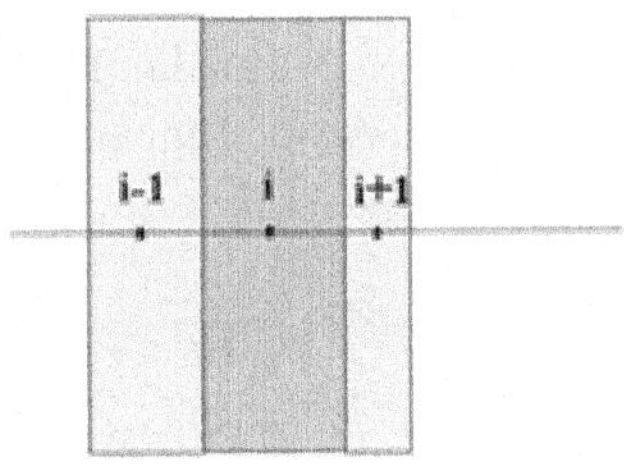

Figure 4.17. *Three adjacent layers*

Layer i receives solar flux from the adjacent layers i-1 and i+1, namely $\phi_{i-1\rightarrow i}$ and $\phi_{i+1\rightarrow i}$.

The "net flux" balances redistributed by the element are expressed as:

$$\begin{cases} \rho_i\phi_{i+1\rightarrow i} + \tau_i\phi_{i-1\rightarrow i} = \phi_{i-1\rightarrow i} \\ \rho_i\phi_{i-1\rightarrow i} + \tau_i\phi_{i+1\rightarrow i} = \phi_{i\rightarrow i+1} \end{cases} \qquad [4.69]$$

The flux absorbed by layer i is:

$$\phi_{abs,i} = \alpha_i(\phi_{i-1\rightarrow i} + \phi_{i+1\rightarrow i}) \qquad [4.70]$$

Point 1 is on the internal face of the multilayer and point n is on the external face which is subjected to the solar radiation. The system of equations to be solved is of the [4.69] type using a specific phrase relating to the outer layers.

$$\begin{cases} \rho_1\phi_{2\rightarrow 1} = \phi_{1\rightarrow 2} \\ \rho_n\phi_{n-1\rightarrow n} + \tau_n\phi_{solar} = \phi_{n\rightarrow n-1} \end{cases} \qquad [4.71]$$

System [4.69, i = 2,...n-1, 4.71] is a system comprising 2(n-2)+2 linear equations and as many unknowns, namely of matrix form AX = B, for which resolution is determined using a certain number of available programs, or even a spreadsheet.

EXAMPLE 4.2.– Take a 3 layer composite, such as a tee-shirt ($\rho_1 = 0.56$; $\tau_1 = 0.36$; $\alpha_1 = 0.08$), a summer shirt ($\rho_2 = 0.61$; $\tau_2 = 0.32$; $\alpha_2 = 0.07$), and a light jacket ($\rho_3 = 0.34$; $\tau_3 = 0.34$; $\alpha 3 = 0.32$). Solving the system gives the solar fluxes absorbed by the layer: $\phi_{abs1} = 0.017\phi_{solar}$; $\phi_{abs2} = 0.039\phi_{solar}$; $\phi_{abs3} = 0.419\phi_{solar}$. These absorbed solar fluxes are therefore the source terms in energy equations.

The calculation for infrared flux is similar to that for solar flux, with the specificity $\alpha_i = \varepsilon_i$ and the expression for an infrared emission in $\varepsilon\sigma T^4$, where T is in Kelvin [BEJ 93]:

$$\begin{cases} \rho_i \phi_{i+1\to i} + \tau_i \phi_{i-1\to i} + \varepsilon_i \sigma T_i^4 = \phi_{i\to i+1} \\ \rho_i \phi_{i-1\to i} + \tau_i \phi_{i+1\to i} + \varepsilon_i \sigma T_i^4 = \phi_{i-1\to i} \end{cases} \quad [4.72]$$

The "boundary" equations should take account of the effects of infrared radiation on skin and the environment:

$$\begin{cases} \rho_1 \phi_{2\to 1} + \tau_1 \varepsilon_{skin} \sigma T_{skin}^4 + \varepsilon_1 \sigma T_1^4 = \phi_{1\to 2} \\ \rho_1 \varepsilon_{skin} \sigma T_{skin}^4 + \tau_1 \phi_{2\to 1} + \varepsilon_1 \sigma T_1^4 = \phi_{1\to skin} \end{cases} \quad [4.73]$$

$$\begin{cases} \rho_n \sigma T_{env}^4 + \tau_n \phi_{n-1\to n} + \varepsilon_n \sigma T_n^4 = \phi_{outgoing} \\ \rho_n \phi_{n-1\to n} + \tau_n \sigma T_{env}^4 + \varepsilon_n \sigma T_n^4 = \phi_{n\to n-1} \end{cases} \quad [4.74]$$

The net flux absorbed by layer i takes into account the fluxes sent by the adjacent layers, as in [4.71], relating to deduction of the emitted fluxes:

$$\varphi_{abs,i} = \alpha_i(\phi_{i-1\to i} + \phi_{i+1\to i}) - 2\varepsilon_i \sigma T_i^4 \quad [4.75]$$

These equations make use of the local temperature (in Kelvin). It will therefore be necessary to consider an iteration. T_{env} is the equivalent environmental temperature. Using the form factor method, it is possible to take into account the heterogeneities of environmental temperature (in the case of fire, for instance).

4.12. Second example: firefighters' clothing

Consider a sample of firefighters' clothing in a fire situation. The vest and trousers are comprised of multiple fire layers which are resistant to fire, heat and water. Let us assume a modern day type of 3 layer clothing: a woven exterior layer of aramid fibers (Nomex®, Kevlar®, Kermel®), an impermeable and breathing middle layer and an interior unwoven thermal lining layer of aramid, which isolates thermally. To this, let us add a tee-shirt layer directly onto skin. We see the infrared coefficients of each layer in the following table.

Layer	1 (interior)	2	3	4 (exterior)
thickness(mm)	0.4	0.4	0.3	0.5
ρ_{IR}	0.10	0.20	0.20	0.40
τ_{IR}	0.40	0.20	0.25	0.15
$\varepsilon_{IR} = \alpha_{IR}$	0.50	0.45	0.55	0.45

Table 4.4. *Data for firefighter's clothing*

We note that these coefficients, which are hard to obtain, should represent an average over a bandwidth corresponding to the temperature of the fire.

In this context where radiation is the essential factor, it is not useful to consider supplementary layers of air, and the same value for conductivity can be used for all layers $\lambda_{eff} = 0.04$ W/mK.

The fixed limit conditions are: $\varepsilon_{skin} = 0.95$; skin temperature $T_p = 37°C$; internal exchange coefficient $h_i = 20$ W/m²K; external temperature $T_{air} = 27°C$; external exchange coefficient $h_e = 20$ W/m²K; fire temperature $T_f = 1{,}100$ K; flame shape factor (solid angle beneath which the flame is "seen") F = 0.15. The flux radiated by the flame is deduced from these ($\sigma = 5.67 * 10^{-8}$ W/m²K⁴): $\dot{q}_f = \sigma T_f^4 F = 12{,}452$ W/m². Environmental flux is therefore expressed by $\sigma F T_f^4 + \sigma(1 - F)T_{env}^4$, with the temperature T_{env} taking into account for form factors for the building and the sky. The external layer is furthermore wetted through aspersion and consequently gives rise to evaporation. We will not consider vapor diffusion through this multilayer, which is a secondary phenomenon in this example, but infrared and conductive/convective transfers (Figure 4.18) [MYH 00].

The equations to be solved in the thermal (stationary) equilibrium situation are as follow: [4.72], [4.73], [4.74] and equation [4.43] degenerated into:

$$\frac{\partial}{\partial x}\left(\lambda_{eff}\frac{\partial T}{\partial x}\right) + \dot{q}_{ray} - \dot{q}_{evap} = 0 \qquad [4.76]$$

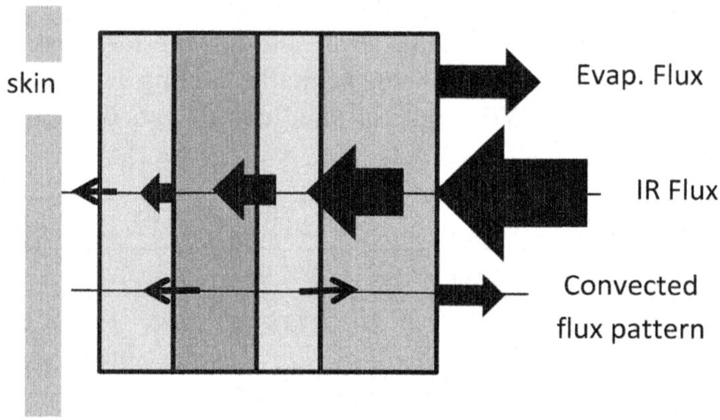

Figure 4.18. *Flux through the firefighter multilayer*

Evaporative flux, which is a source term, is relative to the external layer and we can simplify the problem a little by assuming that its calculation is carried out over a completely humid surface which is subjected to a boundary layer (k = 0.02 m/s) :

$$\dot{q}_{evap} = k(\rho_{vsat}(T_{surf}) - \rho_{vext})\Delta h_v \qquad [4.77]$$

In order to numerically resolve the coupled equations for radiation and conduction, it is convenient to consider each layer as a thermal "entity" with a "node" at the center and a temperature at the node. The system of radiation equations is thus written for i = 1 to i = 4, based on 6 unknowns $\phi_{1\to2}$; $\phi_{2\to1}$; $\phi_{2\to3}$; $\phi_{3\to2}$; $\phi_{3\to4}$; $\phi_{4\to3}$ for 6 equations, the external and internal outgoing will be deduced from this. The discretized conduction equations which assure flux conservation are written:

$$\frac{T_i - T_{i-1}}{R_{i-1,i}} = \frac{T_{i+1} - T_i}{R_{i,i+1}} + \varphi_{abs,i}\, ; i = 2.3 \qquad [4.78]$$

where $R_{i,i+1} = \frac{x_{i+1} - x_i}{\lambda_{eff}}$, the x-axis measured from inside to out.

For the boundary elements, assuming a similar formulation of external exchanges:

$$h_{int}(T_1 - T_p) = \frac{T_2 - T_1}{R_{1,2}} + \varphi_{abs,1} \quad [4.79]$$

$$\frac{T_4 - T_3}{R_{3,4}} = \varphi_{abs,4} - h_{ext}(T_4 - T_{air}) - \dot{q}_{evap} \quad [4.80]$$

The evaporative flux expressed in [4.77] therefore occurs at layer 4 ($T_{surf} = T_4$). Function $\rho_{vsat}(T)$ is given by the humid air equations (Appendix 2). Equations [4.78]–[4.80] form a system of four equations with four unknowns. The two systems of coupled linear equations, 6 of radiation, then 4 of convection, are of the matrix form AX = B and can classically be resolved using "turnkey" software or even a spreadsheet which can calculate the inverse matrix and the product of two matrices. Use of a spreadsheet makes it easy to derive an iteration from the coupling.

Table 4.5 summarizes the calculation results.

T_1 (internal)	T_2	T_3	T_4 (external)	$\Phi_{skin\ net}$	$\dot{q}_{evap}$
58°C	65.9°C	68.2°C	59.4°C	216 W/m^2	5,928 W/m^2

Table 4.5. *Temperature and flux into firefighter's clothing*

It is necessary to take note of the significance of evaporative flux in the cooling of the external clothing layer. The net flux received by skin is the difference between the flux $\phi_{1\rightarrow skin}$ and the flux emitted by skin. This model may enable the importance of the parameters involved in thermal balancing (ρ, τ, ε, λ_{eff}, e_p) to be tested.

4.13. Traditional warm weather clothing

Traditional clothing in warm and dry climates (such as Saharan Africa and Sahel) is generally composed of multiple layers of cotton, presenting a fairly substantial thickness. The requirement is to protect the body from the scorching sun, therefore to reflect and filter solar radiation and to also restrict conductive transfer, which seems logical when air temperature is equal to or even greater than that of skin. How, therefore, is the thermal equilibrium of the body effected? By evaporation at skin level, which may be very significant without causing any apparent run-off, and vapor diffusion through the clothing. Might its composition be optimum so as to limit this evaporation? What type of cotton textile should be chosen and how many layers should be considered?

The laboratory-measured solar reflection and transmission coefficients [SAR 94, BEN 90] (manufacturers ignore these coefficients) present a wide dispersion of values which are in principle unclassifiable through matter, thickness and color. Coefficient ρ_{sol} is generally positioned in the interval (0.3; 0.6) and coefficient τ_{sol} in (0.25; 0.4). Examining the nature of the fabric more closely, we observe that the first factor which determines solar reflection is surface "roughness", caused by yarn fineness, its texturing, the geometric constitution of the armoring and the tension of the woven yarns. The closer the surface is brought to the smooth plane, the greater the reflection. Conversely, knitwear or velvet are solar radiation "traps" and reflect little. We will analyze the different relevant fluxes by way of example.

Consider therefore a possible traditional cotton woven material, of thickness 0.5 mm, with solar coefficients: $\rho_{sol} = 0.55$; $\tau_{sol} = 0.37$; $\alpha_{sol} = 0.08$. Color appears to be an unquantified secondary factor. If we compose a garment using a certain number of identical layers, the global behavior under solar radiation will be deduced from equations [4.68]. For example, for 4 identical superposed layers, then by twice doubling $\rho_{sol} = 0.702$; $\tau_{sol} = 0.068$; $\alpha_{sol} = 0.23$. The increase in solar reflection is considerable. Data from infrared radiation will be estimated for a layer: $\rho_{IR} = 0.10$; $\tau_{IR} = 0.40$; $\alpha_{IR} = 0.50$, and for 4 identical superposed layers: $\rho_{IR} = 0.119$; $\tau_{IR} = 0.027$; $\alpha_{IR} = 0.854$.

The following is the possible meteorological data for Egypt in the hot season: air temperature $T_{air} = 40°C$; relative humidity $\varphi = 0.05$; sun's height at latitude 30° in summer at around midday $h_S = 80°$; the solar flux from blue sky solar radiation, global (direct plus diffuse) $G_o = 1{,}000$ W/m², and direct $B_o = 750$ W/m². The reflection coefficient from the sun on the sand (albedo) is fixed by $\rho_{sol} = 0.4$ and solar temperature $T_{sol} = 50°C$. It only remains to set an equivalent atmospheric temperature for infrared radiation. This is based on the thickness and purity of the atmospheric layer, which reduces with altitude. Away from towns we can set $T_{atm} = 25°C$.

In order to fix an order of magnitude for solar flux for a standing individual we can resume the solar flux integration over a vertical cylinder of height H and diameter D (equation [1.40]) by dividing by surface πDH, then for a flux per square meter:

$$\varphi_{sol} = \frac{B_O}{\pi\, tgh_S} + \frac{D_O + \rho_{sol} G_O}{2} \qquad [4.81]$$

With the previous data, the mean solar flux striking the individual is thus: $\varphi_{sol} = 389.3$ W/m².

Let us now consider heat and vapor transfers through the multilayer composed of 4 layers of cotton (thickness e_p = 0.5 mm), sandwiched between layers of air, simulating the arrangement of clothing (thickness e_a = 1 mm), or a total thickness of e_t of 5 mm. The layers of air have a role in conductive heat and vapor transfer, but not in radiation. Assume steady state and in principle an absence of liquid condensation in the cotton due to the meteorological environment. Simulating moderate activity does not lead to consideration of a filtration velocity toward the multilayer.

Calculation of the transfers is carried out in four stages: solar radiation transmission, heat diffusion, infrared radiation and transmission vapor diffusion (Figure 4.19). Fixing a saturated skin temperature uncouples the diffusive heat and vapor transfers. Infrared radiation and heat diffusion interact and a correction loop as necessary.

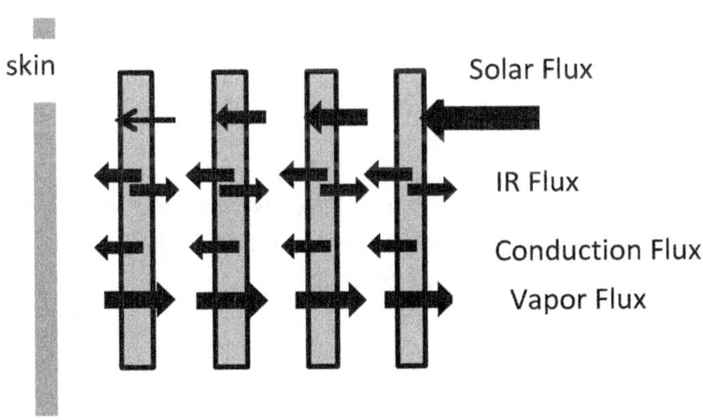

Figure 4.19. *Scheme of fluxes through the solar protection multilayer*

Solar radiation processing takes place by solving the system of equations [4.69]–[4.71] for the 4 identical layers which will not be further broken down. The equation for the internal layer, close to the skin, should be completed in order to take account of a reflection on the skin. Indeed, skin, which reacts differently for solar and infrared radiation, has a solar absorption coefficient α_p of 0.6 for white skin and 0.8 for dark skin [KER 72]. Solar radiation from the first layer (internal, denoted by 1) onto the second (denoted by 2) includes a supplementary term for the flux transferred to the skin, reflected onto the skin and transferred through the first layer:

$$\phi_{1\to 2} = \rho\phi_{2\to 1} + (1-\alpha_p)\tau^2\phi_{2\to 1} \quad [4.82]$$

The system of six equations and six unknowns is easily solved, for example on a spreadsheet (using inversion or products of matrices), and for a solar flux of 389.3 W/m² we obtain the absorbed solar fluxes per layer of cotton (from inside to out): $\varphi_{abs,1}$ = 6.1 W/m²; $\varphi_{abs,2}$ = 14.7 W/m²; $\varphi_{abs,3}$ = 26.1 W/m²; $\varphi_{abs,4}$ = 31.1 W/m² and 16.9 W/m² absorbed by the skin (α_p = 0.6).

Heat and mass diffusion is expressed by the system of heat and mass conservation equations [4.40]–[4.43] degenerated into:

$$\frac{\partial}{\partial x}\left(D_{eff}\frac{\partial \rho_v^i}{\partial x}\right) = 0\,;\ \frac{\partial}{\partial x}\left(\lambda_{eff}\frac{\partial T}{\partial x}\right) + \dot{q}_{rad} = 0 \quad [4.83]$$

The α coefficients from [4.40] have been set as α_1 = 1, α_2 = 0. Vapor diffusion will be dealt with later. The breakdown into basic slices for the calculation is limited to four slices of cotton (of thickness e_p) separated from three layers of air (thickness e_a).

The resolution method selected here is not the nodal method used previously, but an integration of energy equation [4.83], imposing a uniform radiation absorption in each layer of cotton:

$$\varphi(x+e_p) = \varphi(x) + \varphi_{abs}\,;\ \varphi = -\lambda_{eff}\frac{dT}{dx}\,;\ \varphi_{abs} = \dot{q}_{rad}e_p \quad [4.84]$$

$$\lambda_{eff}\big(T(x+e_p) - T(x)\big) = -\varphi(x)e_p - \varphi_{abs}e_p/2 \quad [4.85]$$

For a layer of air we put φ_{abs} = 0 with a very similar conductivity. These equations are written for the four layers of cotton and three of air, namely 14 equations to which we need to add two equations for the limit conditions on the internal (1) and external (2) faces involving the convective exchange coefficients h_i, h_e:

$$\varphi_2 = h_e(T_2 - T_{air});\ \varphi_1 = -h_i(T_1 - T_{skin}) \qquad [4.86]$$

Resolution of this system presents no difficulties and by combining linear equations by reduction we obtain, for example, an equation with one unknown ($\lambda = \lambda_{eff}$):

$$\left[1 + 3\left(1 + \frac{e_a}{e_p}\right) + \frac{\lambda}{h_i e_p} + \frac{\lambda}{h_e e_p}\right]\varphi_1 = -\lambda\frac{T_a - T_p}{e_p} - \left(\frac{1}{2} + \frac{\lambda}{h_e e_p}\right)$$
$$\Sigma\,\varphi_{abs} - (1 + \tfrac{e_a}{e_p})(3\,\varphi_{abs1} + 2\varphi_{abs2} + \varphi_{abs3}) \qquad [4.87]$$

Deduction of the other fluxes and temperatures follow on from each other. It only remains to set the remaining data: h_i = 5 W/m²K; h_e = 20 W/m²K; λ = 0.04 W/mK.

With a first distribution of temperatures being known, the calculation of the infrared exchanges between cotton layers, the exterior and skin may be driven by considering a temperature per layer of cotton which will be the average between those of the two faces which constitute it (equations [4.72]–[4.74]). The temperature T_{env} of [4.74] is obtained from the distribution of external infrared radiation between the sun and sky, with a form factor of ½ for each (vertical cylinder):

$$T_{env}^4 = \frac{1}{2}\left(T_{sky}^4 + T_{sun}^4\right) \qquad [4.88]$$

After several iterations between the "infrared and diffusive systems", the calculation gives us:

Cotton layers	$\varphi_{sol,abs}$ (W/m²)	$\varphi_{IR,\,abs}$(W/m²)	T(°C)
skin	16.9	32.9	35
1	6.1	-32.7	37.0
2	14.7	-17.1	38.3
3	26.1	-20.8	39.7
4	31.1	-17.1	39.9

Table 4.6. *Temperature and flux into solar protection multilayer*

The calculation for diffused vapor flux is thus very simple. Using the vapor diffusion equation in [4.83] and the following limit conditions:

$$-\left(D_{eff}\frac{\partial \rho_v^i}{\partial x}\right)_2 = k_e\left(\rho_v^i(2) - \rho_{v,ext}\right);\ \left(D_{eff}\frac{\partial \rho_v^i}{\partial x}\right)_1 = k_i\left(\rho_v^i(1) - \rho_{v,skin}\right) \qquad [4.89]$$

The classic calculation provides, based on a vapor diffusion coefficient D_{eff} which is identical for the air and cotton layers, the constant vapor flux through multilayer φ_{vap}:

$$\rho_{v,skin} - \rho_{v,ext} = \varphi_{vap}\left(\frac{1}{k_i} + \frac{e_t}{D_{eff}} + \frac{1}{k_e}\right) \qquad [4.90]$$

Let us assume the following numeric values: $k_i = 5*10^{-3}$ m/s, $k_e = 20*10^{-3}$ m/s, $T_{skin} = 35$°C, $e_t = 5$ mm, $D_{eff} = 2*10^{-5}$ m²/s, $\varphi_{humid\ air} = 0.05$, $T_{air} = 40$°C and $T_{skin} = 35$°C. From the humid air data (Appendix 2) $\rho_{vsat}(35°C) = 3.95*10^{-2}$ kg/m³; $\rho_v(40°C, 5\%) = 2.55$ kg/m³, then $\varphi_{vap} = 9.87*10^{-5}$ kg/m²s and an evaporation capacity $P_{evap} = \varphi_{vap}\Delta h_v$; $\Delta h_v(35°C) = 2.57*10^6$ J/kg, i.e. $P_{evap} = 253$ W/m².

For the skin side balance, we therefore have: 49.8 W/m² of solar and infrared contribution and 253 W of losses through evaporation. This evaporative flux is considerable and without doubt the body will react by decreasing its wet skin area, but predicting this would mean a coupling with the human physiological system.

APPENDICES

Appendix 1

Heat and Mass Transfer Toolkit

The continuous variables in a Newtonian viscous fluid medium, based on space and time, are velocity $\vec{V}$, temperature T, density ρ or ρ_v (vapor), enthalpy h, pressure p, etc. The formulation of the conservation laws at low velocities and moderate temperature gradients (constant μ, λ, c_p, etc.) is given below. Thermophysical quantities e, u, h, and c_p are given per unit of mass.

A1.1. Global mass conservation

$$\frac{\partial \rho}{\partial t} + div\, \rho\vec{V} = 0\,; \frac{\partial \rho}{\partial t} + \vec{\nabla}.\rho\vec{V} = 0$$

The second term uses the nabla operator of components in a rectangular reference x, y, z : $\frac{\partial}{\partial x}, \frac{\partial}{\partial y}, \frac{\partial}{\partial z}$. The specific derivative notation is useful: $\frac{D}{Dt} = \frac{\partial}{\partial t} + \vec{V}.\vec{\nabla}$ and enables the previous equation to be rewritten in the form:

$$\frac{D\rho}{Dt} + \rho\vec{\nabla}.\vec{V} = 0$$

A1.2. Momentum conservation

$$\rho\frac{D\vec{V}}{Dt} = -\vec{\nabla}p + \rho\vec{g} + \mu\vec{\nabla}(\vec{\nabla}.\vec{V})$$

Still in reference x, y, z, the component over x of this equation for component u of the velocity vector is as follows:

$$\rho\left(\frac{\partial u}{\partial t}+\vec{V}.\vec{\nabla}u\right)=-\frac{\partial p}{\partial x}+\rho\vec{g}.\vec{x}+\mu\vec{\nabla}.\overrightarrow{\nabla u}$$

This equation is the basis for the study of boundary layers, whether from forced or natural convection, and a number of simplifications will be derived. For natural convection, the gravity term will be the engine for the motion and the density ρ will be based on the temperature for this term, whereas it is taken as a constant in the inertia term (left-hand side).

A1.3. Energy conservation or the first principle of thermodynamics

$$\rho\frac{Dh}{Dt}=\frac{Dp}{Dt}+\mu\phi+\vec{\nabla}.(\lambda\overrightarrow{\nabla T})$$

In the context of applications considered here, the energy from compression and expansion, as well as the viscous dissipation ϕ, is negligible. The enthalpy for a perfect gas and a liquid is written $Dh=c_pDT$, which provides a simplified equation for convection problems:

$$\rho c_p\left(\frac{\partial T}{\partial t}+\vec{V}.\vec{\nabla}T\right)=\vec{\nabla}.(\lambda\vec{\nabla}T)$$

A1.4. Vapor mass conservation

Evaporation and condensation involve low amounts of water vapor contained in air ($\rho_v/\rho \approx 10^{-3}, 10^{-2}$) and water vapor mass, which expresses vapor diffusion in air, should be considered. Let us assume that liquid and vapor state changes take place on the solid walls or liquid surfaces, i.e. that problems from misting are not considered, and this conservation is written as:

$$\left(\frac{\partial\rho_v}{\partial t}+\vec{V}.\vec{\nabla}\rho_v\right)=\vec{\nabla}.(D_v\vec{\nabla}\rho_v)$$

Assuming that the coefficient for vapor diffusion in air D_v is constant and in reference x, y, z, this equation is in the same form as the energy equation, which will be useful when formulating the analogies between thermal and mass (humidity) boundary layers.

These equations are accompanied by fluid equations of state, in general air and water vapor contained in air, expressing partial pressure and the perfect gases equation for air and vapor (Appendix 2):

$$\begin{cases} p = p_a + p_v \,;\ \rho = \rho_a + \rho_v \\ p_a = \rho_a \dfrac{R}{Mol_a} T \,;\ p_v = \rho_v \dfrac{R}{Mol_v} T \end{cases}$$

A1.5. Energy balance in a finite volume (material system)

Consider an "open" volume defined by a domain D which allows mass flows in or out of sections σ_i. The balance equation is written as:

$$\begin{cases} \dfrac{dE}{dt} + h_i \dot{m}_i = \dot{Q} \\ E = \int_{\mathcal{D}} \rho e d\mathcal{V};\ e = u + \dfrac{V^2}{2} + gz \\ h = e + p/\rho \end{cases}$$

The heat flux $\dot{Q}$ "received" by the system may be a flux of radiation, combustion or convection at the domain boundary. For the context considered, the potential and kinetic energy is negligible, the perfect gases satisfying $e = u = c_v T$ (2 = sign), and liquids or solids satisfying $c_p = c_v$.

A1.6. Energy balance on surface subject to evaporation or condensation and radiation

Consider a liquid film on a solid or free liquid surface. A normal is directed toward the external gaseous medium (n_{ext}) another toward the internal liquid or solid medium (n_{int}). The surface emits a net infrared radiation flux $\dot{q}_{rad}$ and the flux balance per unit of area is written as:

$$-D_v \frac{\partial \rho_v}{\partial n_{ext}} \Delta h_v = -\dot{q}_{rad} + \lambda_{ext} \frac{\partial T}{\partial n_{ext}} + \lambda_{int} \frac{\partial T}{\partial n_{int}}$$

By introducing exchange coefficients, if they occur, the heat and mass (evaporation/condensation) fluxes are expressed as:

$$-\lambda_{ext}\frac{\partial T}{\partial n_{ext}} = h_{ext}(T_p - T_{ext});\ -\lambda_{int}\frac{\partial T}{\partial n_{int}} = h_{int}(T_p - T_{int});$$

$$-D_v\frac{\partial \rho_v}{\partial n_{ext}} = k(\rho_{vp} - \rho_{vext})$$

A1.7. Some heat exchange correlations for air and water [BEJ 93]

A1.7.1. *Natural convection*

Boundary layer on a vertical flat plate of height H (Churchill and Chu correlation):

$$\overline{Nu_H} = \left\{0.825 + 0.387Ra_H^{1/6}\left[1 + (a/Pr)^{9/16}\right]^{-8/27}\right\}^2$$

$$\overline{Nu_H} = \frac{\bar{h}H}{\lambda} = \frac{\bar{q}}{T_p - T_o}\frac{H}{\lambda};\ Ra_H = \frac{g\beta(T_p - T_o)H^3}{\alpha\nu}$$

where $T_p = cst \rightarrow a=0.492$, $Ra_H<10^{12}$; constant flux$\rightarrow a=0.437$.

Boundary layer on a heated horizontal cylinder with T_p constant (Churchill and Chu):

$$\overline{Nu_D} = \left\{0.6 + 0.387Ra_D^{1/6}\left[1 + (0.559/Pr)^{9/16}\right]^{-8/27}\right\}^2$$

with condition: $Ra_D<10^{12}$. Note that turbulent transition takes place at around $Ra_D=10^9$.

Boundary layer on a vertical cylinder of height H (Lienhard):

$$\overline{Nu_H} = 0.52Ra_H^{1/4}$$

For a vertical cylinder with diameter equal to the height, Sparrow gives:

$$\overline{Nu_D} = 0.775Ra_D^{0.208};\ Ra_D > 1.4 * 10^4$$

Boundary layer on an isothermal sphere (Churchill):

$$\overline{Nu_D} = 2 + 0.589Ra_D^{1/4}\left[1 + (0.469/Pr)^{9/16}\right]^{-4/9}$$

where $Ra_D < 10^{11}$.

Boundary layer on a horizontal plate of length L with various isothermal heating or cooling conditions. Heated upper surface (heated plume which rises) or cooled lower surface (cold plume which falls) (Lloyd):

$$\overline{Nu_L} = 0.54Ra_L^{1/4};\ 10^4 < Ra_L < 10^7$$

$$\overline{Nu_L} = 0.15Ra_L^{1/3}\ ;\ 10^7 < Ra_L < 10^9$$

A heated lower or cold upper surface determines a different flow:

$$\overline{Nu_L} = 0.27Ra_L^{1/4};\ 10^5 < Ra_L < 10^{10}$$

Exchanges between two horizontal plates, distance H apart, the upper plate cooled to T_f and the lower plate heated to T_c. Natural convection turbulence (Bénard) is established from $Ra_H > 1708$ (Arnold):

$$\begin{cases} \overline{Nu_H} = 0.069Ra_H^{1/3}Pr^{0.074}\ ;\ Ra_H = \frac{g\beta}{\alpha\nu}(T_c - T_f)H^3 \\ \overline{Nu_H} = \frac{\bar{q}H}{\lambda\Delta T}\ ; 3 * 10^5 < Ra_H < 7 * 10^9 \end{cases}$$

A1.7.2. Forced convection

Laminar boundary layer on flat plate at constant temperature (Blasius):

$$\overline{Nu_x} = 0.664Pr^{1/3}Re_x^{1/2};\ Re_x = \frac{Vx}{\nu}$$

The Nusselt number is integrated from 0 to x.

Turbulent boundary layer on flat plate following a transition to $Re_x \cong 5*10^5$:

$$\overline{Nu_L} = 0.037Pr^{1/3}\left(Re_L^{4/5} - 23550\right)$$

The Nusselt number is integrated from 0 to L and $5*10^5<Re_L<10^8$.

Boundary layer on cylinder with diameter D, perpendicular to the flow velocity:

The most classical:

$$\overline{Nu_D} = 0.26Pr^{0.37}Re_D^{0.6};\ 10^3 < Re_D < 2*10^5$$

$$\overline{Nu_D} = 0.023Pr^{0,4}Re_D^{0.8};\ 2*10^5 < Re_D < 10^7$$

One more general (Churchill, Bernstein):

$$\overline{Nu_D} = 0.3 + 0.62Pr^{1/3}Re_D^{1/2}\left[1 + (0.4/Pr)^{2/3}\right]^{-1/4}\left[1 + \left(\frac{Re_D}{282{,}000}\right)^{5/8}\right]^{4/5}$$

The values for the physical parameters are calculated for the average temperature of the wall and the upstream fluid. This latter correlation may equally be applied for a constant heat flux condition.

Boundary layer on a sphere with diameter D in a fluid current (Whitaker):

$$\overline{Nu_D} = 2 + Pr^{0.4}\left(0.4Re_D^{1/2} + 0.06Re_D^{2/3}\right);\ Re_D < 7.6*10^4$$

may be applied for a temperature or constant flux condition at the wall.

Appendix 2

Humid Air

Humid air, from a particular fluid, a homogeneous volume, or a section of pipe, for example, is typically a mixture of dry air (a) and water vapor (v) at normal atmospheric pressure (p = 101300 Pa). At other pressures and for other gaseous components, a reformulation of equations is required [FOH 10].

The laws of mixed gas states (the Dalton's laws) link partial pressures p_a, p_v, densities ρ_a, ρ_v and temperature T:

$$p_a = \rho_a r_a T;\ p_v = \rho_v r_v T;\ p_a + p_v = p$$

T expressed with Kelvin scale, r_a and r_v are the constants, r = R/Mol, where R is ideal gas constant, Mol is molar mass.

The quantity of water vapor in a given volume is limited according to the saturation state which corresponds to the appearance of liquid water. The corresponding partial pressure is said to be from saturation and is only based on temperature:

$$p_v max = p_{sat}(T)$$

One measure of the quantity of water vapor in air is moisture content:

$$\phi = \frac{p_v}{p_{vsat}(T)} = \frac{\varrho_v}{\varrho_{vsat}(T)}$$

ϕ is, therefore, a value between 0 and 1, or usually more meaningfully, between 0 and 100%.

Another measure of water in a volume of humid air is absolute humidity or weight:

$$W = \frac{\varrho_v}{\varrho_a}$$

This ratio is dimensionless but is often noted by reference to its definition as the weight of water, divided by the weight of dry air (or g/kg dry air, for instance). The range of values for W is mostly between 10^{-3} and 10^{-2}.

W and ϕ are linked by the relationships above:

$$W = a\frac{\phi p_{vsat}(T)}{p-\phi p_{vsat}(T)};\, a = \frac{r_a}{r_v}$$

From the definition for W, the enthalpy of the mixture between air and water vapor is expressed per kg of dry air:

$$h = h_a + W h_v$$

Assuming perfect gas, air and water vapor satisfy $dh = c_p dT$, where values for the specific heat c_p are constant in the field of study. Furthermore, we set reference $h = 0$ at 0°C, which simplifies the expression without affecting the balances which involve enthalpy differences.

Here, we need to expand the field of humid air to the occurrence of liquid water (mist) or solid water (snow) within the basic volume considered. State changes between solid and liquid or between liquid and gas give rise to significant variations in enthalpy (latent heat) that should be stated as:

$$\Delta h_{vap}(T) = h_{vap} - h_{liq}\,;\, \Delta h_{liq}(T) = h_{liq} - h_{sol}$$

We note Δh_v (0°C) = Δh_{v0} and Δh_{liq} (0°C) = Δh_{g0}.

The measure of the quantity of water remains W and the vapor element of liquid water or ice is denoted by W_{sat}. The enthalpy reference of the mix in liquid or ice is, therefore, defined as 0°C in liquid state. The expressions for enthalpy in three cases vapor, vapor + liquid and vapor + ice are as follows (T in °C):

$$\begin{cases} W < W_{sat}(T); h = c_{pa}T + W(\Delta h_{v0} + c_{pv}T) \\ W > W_{sat}(T); T > 0;\ h = c_{pa}T + W_{sat}(T)(\Delta h_{v0} + c_{pv}T) + (W - W_{sat})c_{liq}T \\ W > W_{sat}(T); T < 0; h = c_{pa}T + W_{sat}(T)(\Delta h_{v0} + c_{pv}T) - (W - W_{sat})(\Delta h_{g0} - c_{liq}T) \end{cases}$$

These equations enable the parameters for humid air such as T, ϕ, p_v, ρ_a, W and h to be calculated from two values entered, most often T and ϕ. A scientific pocket calculator does the job. Humid air charts (which are psychometric charts) are useful for following the development of an air conditioning technology loop. These Mollier, or modified Mollier, diagrams depict graphs for constant values T, W, h, ϕ and $1/\rho_a$.

Satisfactory values at a temperatures below 50°C are:

p = 101300 Pa; c_{pa} = 1.006 kJ/kgK; c_{pv} = 1.860 kJ/kgK; c_{liq} = 4.180 kJ/kgK; c_g = 2.110 kJ/kgK;

r_a = 0.287 kJ/kgK; r_v = 0.461 kJ/kgK; a = 0.622; Δh_{v0} = 2501.6 kJ/kg; Δh_{g0} = 333 kJ/kg

The equation for $p_{vsat}(T)$ can be determined from various expressions with varying degrees of accuracy based on ranges of temperatures. For T<50°C, the following is an expression which is quite easy to calculate [CAD 06]:

$$log_{10}p_{vsat}(T) = 2.7858 + 7.5 * \frac{T}{237.3+T}$$

with T in °C and p in Pascal.

Appendix 3

Solar Flux

The calculation of solar flux on a surface of given orientation involves geometric data (surface position), time (date and hour), site (latitude, pollution, state of the ground, etc.), as well as statistical meteorological data (such as cloud coverage). The two most common types of problems presented are calculating a likely average energy over a given period (1 month, say), or calculating a maximum possible energy due in good weather at a given time. The data collected by weather stations include temperature and humidity of air, global solar radiation and diffuse radiation over a horizontal surface, the insolation fraction (the ratio of the number of hours of sunshine). From the data recorded, it will be possible to extract an instantaneous solar flux (W/m^2), or an average flux, for example monthly (kJ/m^2 or KWh/m^2) [DUF 74].

We must, therefore, consider a direct solar flux (denoted by B), a diffuse solar flux which is diffused by reflection from the ground and sky (D) and a global flux (G), which is the sum of B and D. These fluxes are based on the position of the sun determined by its azimuth (a_{sol}) and its height, i.e. the view angle relative to the horizontal (h), as well as the position of the considered surface determined by its azimuth, a_{surf}, and its inclination relative to the horizontal plane (i). The azimuths are measured relative to the south (position zero), with negatives to the east and positives to the west.

With a co-ordinates system x, y, z where, for example, x is to the south and z is to the vertical, the coordinates of vector $\vec{n}$ normal to the surface facing the direct solar radiation are ($\sin i.\cos a_{surf}$, $-\sin i.\sin a_{surf}$, $\cos i$). The coordinates of the direct solar flux vector $\vec{I}$ (irradiation) directed toward the sun are ($\cosh.\cos a_{sol}$, $-\cosh.\sin a_{sol}$, $\sinh$). The direct solar flux on the surface is $\vec{I}.\vec{n}$ = I(cosh.sini.cos(a_{sol}-a_{surf})+sinh.cosi). The direct flux on a horizontal surface is, therefore, B_0 = I.sinh, and by eliminating I, the direct flux on the surface becomes:

$$B = B_0 R_b;\ R_b = \frac{cosh.sini.\cos(a_{sol} - a_{surf}) + sinh.cosi}{sinh}$$

Diffuse radiation is the sum of the diffusion originating from the ground and sky. For a surface inclined at angle i relative to the horizontal, the solid angle below which it "sees" the sky is (1+cosi)/2 and the angle below which it "sees" the ground is (1-cosi)/2. We must furthermore take account of the albedo solar reflection of the ground ρ_{sol}, which quantifies the reflected global flux. If D_0 and G_0 = B_0+D_0 denote the radiation measured on the ground, the global and diffuse fluxes on an inclined surface are:

$$G = B_0 R_b + D\ ;\ D = D_0 \frac{1+cosi}{2} + \rho_{sol} G_0 \frac{1-cosi}{2}$$

Solar angles h and a_{sol} are obtained based on position (latitude φ), on the day of year J and on time AH. Converting clock time into AH is carried out as follows: AH = 0 at solar midday, then increases at 15° per hour after midday and decreases by the same before midday. The declination angle δ of the Earth on its elliptical path around the sun is calculated based on the day of year (from 0 to 365). The geometric formulae are written as:

$$\begin{cases} sinh = sin\varphi.sin\delta + cos\varphi.cos\delta.cosAH \\ sina_{sol} = cos\delta.sinAH/cosh \\ cosa_{sol}.cosh = sin\varphi.cos\delta.cosAH - cos\varphi.sin\delta \\ \delta = 23.45° sin\left(360° \frac{284+J}{365}\right) \end{cases}$$

These formulae enable the hours of sunrise and sunset, for example, to be determined, as well as the maximum "height" of the sun h_{max} = π/2-φ+δ.

For a study on a time of year, local data of the form $G_0(\sigma)$, $D_0(\sigma)$, where σ is an insolation fraction between 0 and 1, could be determined. For a study of the maximum clear sky values, some experimental formulae provide G_0 based on the height h of the sun, such as $G_0 = 1210 \exp\left(-\frac{1}{6}\sin h\right)$ W/m², for example, where D_0 = 0.3G_0.

Bibliography

[BAC 91] BACON C.J., BELL S.A., CLULOW E.E. *et al.*, "How much mothers keep their babies warm", *Archives of Disease in Childhood*, vol. 66, pp. 627–632, 1991.

[BEJ 84] BEJAN A., *Convection Heat Transfer*, Wiley, New York, 1984.

[BEJ 93] BEJAN A., *Heat Transfer*, Wiley, New York, 1993.

[BEN 90] BEN AÏM H., Etude des transferts de chaleur sèche au sein et au travers d'une étoffe textile, Caractérisation expérimentale-modélisation, INP Thesis Toulouse, 1990.

[CAB 97] CABANAC M., BONNIOT-CABANAC M.C., "De quoi serait mort le coureur de Marathon?", *Médecine/Sciences*, vol. 13, pp. 838–842, 1997.

[CRA 75] CRANK J., *The Mathematics of Diffusion*, Clarendon Press, Oxford, 1975.

[DUF 74] DUFFIE J.A., BECKMAN W.A., *Solar Energy Process*, Wiley, New York, 1974.

[FAN 70] FANGER P.O., *Thermal Comfort*, MacGraw-Hill, New York, 1970.

[FOH 02] FOHR J.P., COUTON D., TREGUIER G., "Dynamic heat and water transfer through layered fabrics", *Textile Research Journal*, vol. 72, no. 1, pp. 1–12, 2002.

[FOH 10] FOHR J.P., *Thermodynamique des systèmes fluides et des machines thermiques*, Hermes Science Lavoisier, 2010.

[GAG 71] GAGGE A.P., STOLWIJK J.A.J., NISHI Y., "An effective temperature scale based on a simple model of human physiological regulatory response", *ASHRAE Transactions*, vol. 77, pp. 247–262, 1971.

[HAV 90] HAVENITH G., HENS R., LOTENS W.A., "Clothing ventilation, vapour resistance and permeability index: changes due to posture movement and wind", *Ergonomics*, vol. 33, no. 8, pp. 989–1005, 1990.

[HAV 99] HAVENITH G., HOLMER I., DEN HARTOG E.A. *et al.*, "Clothing evaporative heat resistance-proposal for improved representation in standards and models", *Annals of Occupational Hygiene*, vol. 43, no. 5, pp. 339–346, 1999.

[JAR 92] JARDINE D.S., "A mathematical model of life-threatening hyperthermia during infancy", *Journal of Applied Physiology*, vol. 73, no. 1, pp. 329–339, 1992.

[KER 72] KERSLAKE D. MCK., *The Stress of Hot Environments*, Cambridge University Press, 1972.

[MOR 93] MORTON W.E., HEARLE J.W.S., *Physical Properties of Textile Fibres*, 3rd ed., Textile Institute, Manchester, UK, 1993.

[MYH 00] MYHRE L.G., BARKER R.L., MISZKO T. *et al.*, "Effectiveness of breatheable moisture barriers in dissipating metabolic heat in working fire fighters", *9th International Conference on Environmental Ergonomics*, Dortmund, Germany, pp. 349–351, 2000.

[PAT 80] PATANKAR S.V., *Numerical Heat Transfer and Fluid Flow*, Hemisphere Publishing Corporation, McGraw-Hill, New York, 1980.

[SAD 87] SADIK KAKAÇ, RAMESH K.S., WIN AUNG, *Handbook of Single-Phase Convective Heat Transfer*, Chapter 6, Wiley, New York, 1987.

[SAR 94] SARI H., Interface vêtement: échanges hygrothermiques et microclimat sous-vestimentaire, University of Nice Sophia-Antipolis, 1994.

[SCH 60] SCHLICHTING H., *Boundary Layer Theory*, McGraw Hill, New York, 1960.

[STO 70] STOLWIJK J.A.J., "Mathematical model of thermoregulation", in HARDY J.D., GAGGE A.P., STOLWIJK J.A.J. (eds), *Physiological and Behavioral Temperature Regulation*, Charles C. Thomas Publisher, Springfield, IL, pp. 703–721, 1970.

[STO 71] STOLWIJK J.A.J., A mathematical model of physiological temperature regulation in man, NASA Report CR.1855, NASA, Washington, DC, 1971.

[STU 84] STUART I.M., HOLCOMBE B.V., "Heat transfer through fiber beds by radiation with shading and conduction", *Textile Research Journal*, vol. 54, pp. 149–157, 1984.

[SUL 73] SULOK E., JEQUIER E., PROD'HOM L.S., "Thermal balance of the newborn in a heat-gaining environment", *Pediatric Research*, vol. 7, pp. 888–900, 1973.

[WEH 88] WEHNER J., MILLER B., REBENFELD L., "Dynamics of water vapor transmission through fabric barriers", *Textile Research Journal*, vol. 58, pp. 581–592, 1988.

[WHI 98] WHITAKER S., "Coupled transport in multiphase systems: a theory of drying", *Advances in Heat Transfer*, Academic Press, New York, vol. 31, pp. 1–104, 1998.

ISO standards

9237 (1995), Determination of the permeability of fabrics to air.

11092 (2014), Measurement of thermal and water-vapor resistance under steady-state conditions: Sweating guarded hotplate test.

5084 (1996), Determination of thickness of textiles and textile products.

3801 (1977), Textiles – Woven fabrics – Determination of mass per unit length and mass per unit area.

7726 (1998), Ergonomics of the thermal environment: instruments for measuring physical quantities.

7730 (2005), Ergonomics of the thermal environment: analytical determination and interpretation of thermal comfort using calculation of the PMV and PPD indices and local thermal comfort criteria.

9920 (2007), Ergonomics of the thermal environment: estimation of thermal insulation and water vapor resistance of a clothing ensemble.

Index

A, B, C, E

absorption, 9, 113, 114, 130, 131, 140
bound water, 89, 104–111, 119–121, 123, 124
boundary layer, 3–8
buoyancy, 30, 32–34, 36, 37, 39
condensation, 10, 11, 88, 101, 105, 109, 111, 118–121, 124–128, 139
emission (emittance), 9, 62, 131, 134
evaporative resistance, 89, 103, 104
exchange coefficient, 75, 88, 109, 121, 135, 141

F, G, I

Fanger model, 61–69, 74–77
filtration, 100–102, 105, 108, 120–124, 128–130, 139
forced convection, 4, 5, 8, 9, 19, 20, 43–59, 64, 128
Gagge model, 76–80
Grashof number, 6
imbibition, 115–117
inertial, 115
infrared radiation, 21, 65, 101, 124, 131, 134, 138–141
integral
  equation, 34, 39, 40, 46, 48, 49
  method, 26, 43, 45, 47, 48, 54

M, N, P

metabolism, 19, 22, 23, 62, 63, 67, 68, 73, 75, 77–79, 83, 91, 92
multilayer, 99, 122–130, 133, 135, 136, 139, 141, 142
natural convection, 2, 6, 8, 12, 13, 15, 18–20, 25–37, 38, 45, 47, 64, 84
nodal method, 85, 126, 140
Nusselt number, 5, 6, 20, 24, 36, 58, 59, 85, 121
Peclet number, 129, 130
pore, 67, 77, 99, 112, 114, 115
Prandtl number, 5, 15, 28, 32, 56, 59, 85
pressure coefficient, 46, 50, 122

R, S

Rayleigh number, 6, 8, 14, 17, 20, 26, 43
reflection, 9, 21, 130, 131, 132, 138, 140
Reynolds number, 4, 5, 20, 24, 45, 50, 59
Schmidt number, 5, 39, 85

Sherwood number, 5, 42, 85
skin wettedness, 69
solar radiation, 9, 19, 21, 22, 124, 130, 131, 133, 137–140
sorption, 104, 106, 107, 110
Stolwijk model, 81

T, V

temperature profile, 26, 29, 57, 129
thermal
comfort, 61, 68, 70, 72
resistance, 64, 66–68, 87–89, 103, 104
time scale, 105, 111, 118, 124
transition, 6, 8, 24, 26, 44, 68, 73, 74, 100
turbulence, 4, 8, 44, 71, 85
vasoconstriction, 23, 76, 78, 80, 81
vasodilatation, 63, 76, 78, 80, 81, 89, 92
velocity profile, 45, 49–52, 54

Other titles from

iSTE

in

Fluid Mechanics

2014

CALTAGIRONE J.-P.
Discrete Mechanics

PRUD'HOMME Roger
Flows and Chemical Reactions in Heterogeneous Mixtures
Flows and Chemical Reactions in an Electromagnetic Field

TARDU Sedat
Transport and Coherent Structures in Wall Turbulence

2013

BORGHI Roland, ANSELMET Fabien
Turbulent Multiphase Flows with Heat and Mass Transfer

DÉLERY Jean
Three-dimensional Separated Flow Topology: Critical Points, Separation Lines and Vortical Structures

PRUD'HOMME Roger
Flows and Chemical Reactions in Homogeneous Mixtures

2012

BOUTIER Alain
Laser Velocimetry in Fluid Mechanics

BOUTIER Alain
Laser Metrology in Fluid Mechanics: Granulometry, Temperature and Concentration Measurements

PRUD'HOMME Roger
Flows and Chemical Reactions

2011

MORY Mathieu
Fluid Mechanics for Chemical Engineering

TARDU Sedat
Statistical Approach to Wall Turbulence

2010

COLIN Stéphane
Microfluidics

DÉLERY Jean
Handbook of Compressible Aerodynamics

GUINOT Vincent
Waves Propagation in Fluids – 2nd edition

SOULI Mhamed, BENSON J. David
Arbitrary Lagrangian Eulerian and Fluid-structure Interaction

2009

FAVRE-MARINET Michel, TARDU Sedat
Convective Heat Transfer

2008

PEUBE Jean-Louis
Fundamentals of Fluid Mechanics and Transport Phenomena

SCHIESTEL Roland
Modeling and Simulation of Turbulent Flows

CPSIA information can be obtained at www.ICGtesting.com
Printed in the USA
BVOW06*1649031115

425328BV00001B/10/P